NX 12 Tutorial

Online Instructor

Limit of Liability/Disclaimer of Warranty:
The author and publisher make no representations or warranties with respect to the accuracy or completeness of the contents of this work and specifically disclaim all warranties, including without limitation warranties of fitness for a particular purpose. The advice and strategies contained herein may not be suitable for every situation. Neither the publisher nor the author shall be liable for damages arising here from.

Trademarks:
All brand names and product names used in this book are trademarks, registered trademarks, or trade names of their respective holders. The author and publisher are not associated with any product or vendor mentioned in this book.

Contents

Introduction

NX as a topic of learning is vast, and having a wide scope. It is one of the world's most advanced and highly integrated CAD/CAM/CAE product. NX delivers a great value to enterprises of all sizes by covering the entire range of product development. It speeds up the design process by simplifying complex product designs.

This tutorial book provides a systematic approach for users to learn NX 12. It is aimed for those with no previous experience with NX. However, users of previous versions of NX may also find this book useful for them to learn the new enhancements. The user will be guided from starting a NX 12 session to constructing parts, assemblies, and drawings. Each chapter has components explained with the help of various dialogs and screen images.

Scope of this Book

This book is written for students and engineers who are interested to learn NX 12 for designing mechanical components and assemblies, and then generate drawings.

This book provides a systematic approach for learning NX 12. The topics include Getting Started with NX 12, Basic Part Modeling, Constructing Assemblies, Constructing Drawings, Additional Modeling Tools, and Sheet Metal Modeling.

Chapter 1: Introduces NX 12. The user interface, terminology, mouse functions, and shortcut keys are discussed in this chapter.

Chapter 2: Takes you through the creation of your first NX model. You construct simple parts.

Chapter 3: Teaches you to construct assemblies. It explains the Top-down and Bottom-up approaches for designing an assembly. You construct an assembly using the Bottom-up approach.

Chapter 4 teaches you to generate drawings of the models constructed in the earlier chapters. You will also learn to generate exploded views, and part list of an assembly.

Chapter 5: In this chapter, you will learn the tools needed to create 2D sketches.

Chapter 6: In this chapter, you will learn additional modeling tools to construct complex models.

Chapter 7: This chapter helps you to create, edit, and use expressions in your designs.

Chapter 8: introduces you to NX Sheet Metal design. You will construct a sheet metal part using the tools available in the NX Sheet Metal environment.

Chapter 9: teaches you to create an assembly using Top-down design approach.

Chapter 10: teaches you to add dimensions and annotations to your drawings.

Chapter 11: introduces you to Finite Element Analysis.

Chapter 12: teaches you to add Product and Manufacturing Information to 3D models

Chapter 13: teaches you to add materials and scene to models. Also, it helps you to render photorealistic images.

Chapter 1: Getting Started

In this chapter, you will learn some of the most commonly used features of NX. Also, you will learn about the user interface.

In NX, you construct 3D parts and use them to generate 2D drawings and 3D assemblies.

NX is Feature Based. Features are shapes that are combined to build a part. You can modify these shapes individually. For example, the following figure shows a part built using the Extrude and Hole features.

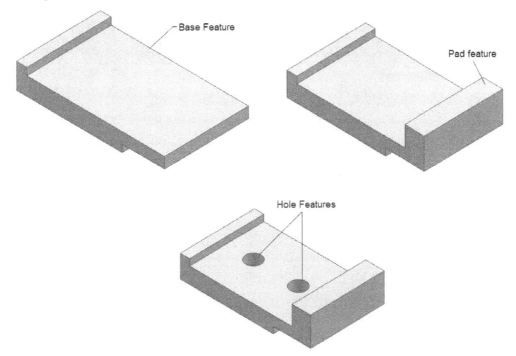

Most of the features are sketch-based. A sketch is a 2D profile and can be extruded, revolved, or swept along a path to construct features.

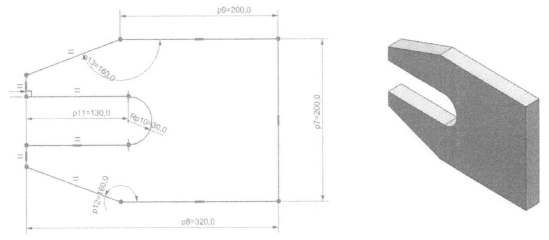

NX is parametric in nature. You can specify standard parameters between the elements of a part. Changing these parameters changes the size and shape of the part. For example, see the design of the body of a flange before and after modifying the parameters of its features.

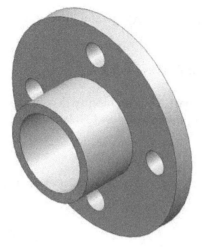 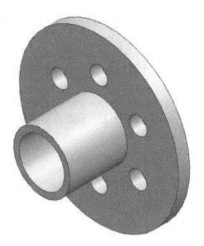

Starting NX

1. Click the Windows button on the taskbar.
2. Scroll down to the **S** section.
3. Click **Siemens NX 12.0 > NX 12.0**.
4. Click **Home** tab > **New** button on the ribbon.
5. On the **New** dialog, click **Templates > Model**.
6. Click the **OK** button.

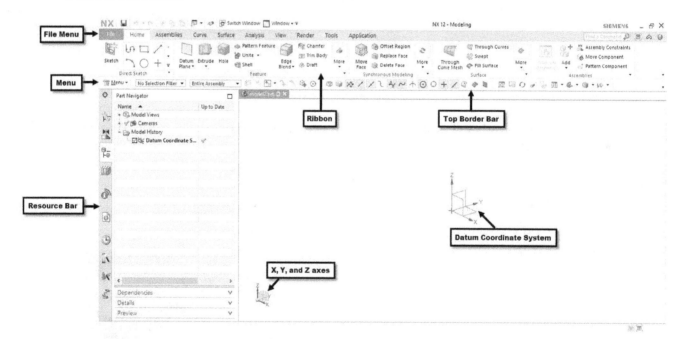

Notice these important features of the NX window.

User Interface

Various components of the user interface are discussed next.

Quick Access Toolbar

This is located at the top left corner of the window. It consists of the commonly used commands such as **Save**, **Undo, Redo, Copy**, and so on.

File Menu

The **File Menu** appears when you click on the **File** icon located at the top left corner of the window. The **File Menu** consists of a list of self-explanatory menus. You can see a list of recently opened documents under the **Recently Opened Parts** section. You can also switch to different applications of NX.

Ribbon

A ribbon is a set of tools, which are used to perform various operations. It is divided into tabs and groups. Various tabs of the ribbon are discussed next.

Home tab

This ribbon tab contains the tools such as **New, Open, and Help,** and so on.

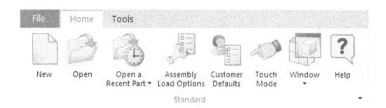

Home tab in the Model template

This ribbon tab contains the tools to construct 3D features.

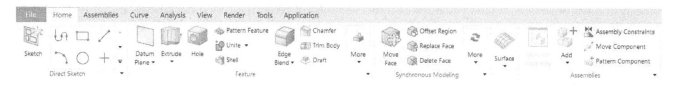

View tab

This ribbon tab contains the tools to modify the display of the model and user interface.

Analysis tab

This ribbon tab has the tools to measure the objects. It also has tools to analyze the draft, curvature, and surface.

Home tab in Sketch Task environment

This ribbon tab contains all the sketch tools. It is available in a separate environment called Sketch Task environment. The Sketch Task environment is activated when you activate a Feature modeling tool and click on a planar face or Datum plane.

Tools tab

This ribbon tab contains the tools to create expressions, part families, movies, fasteners.

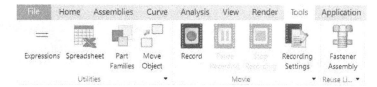

Render tab

This ribbon tab contains the tools to generate photorealistic images.

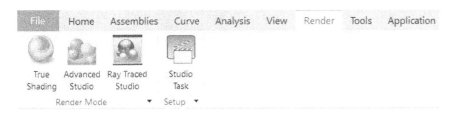

Application tab

This ribbon tab contains the tools to start different applications such as Assemblies, Sheet Metal, Drafting, and so on.

Assemblies tab

This tab contains the tools to construct an assembly. It is available in the **Assembly** and **Model** template.

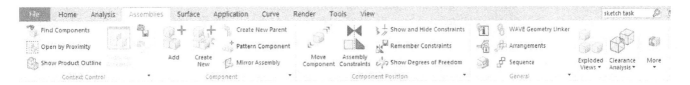

Drafting environment ribbon

In the Drafting Environment, you can generate orthographic views of the 3D model. The ribbon tabs in this environment contain tools to generate 2D drawings.

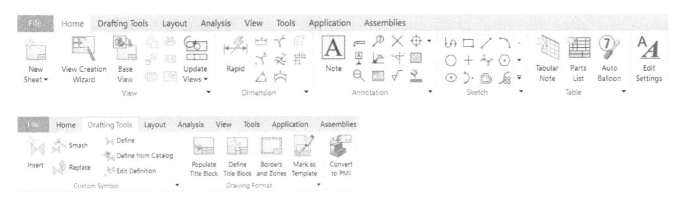

Sheet Metal ribbon

The tools in this ribbon are used to construct sheet metal components.

Some tabs are not visible by default. To display a particular tab, right-click on the ribbon and select it from the list displayed.

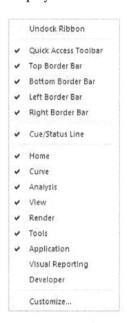

You can also add a ribbon tab by opening the **Customize** dialog. Click the down arrow located at the bottom right corner of the ribbon, and select **Customize**. On the **Customize** dialog, click on different tabs and select/deselect the options.

Ribbon Groups and More Galleries

The tools on a ribbon are arranged in various groups depending upon their use. Each group has a **More Gallery**, which contain additional tools.

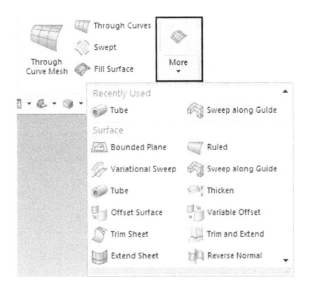

You can add more tools to a ribbon group by clicking the arrow located at the bottom right corner of a group.

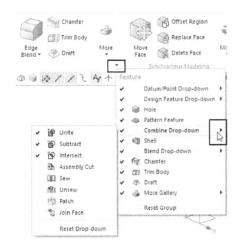

Top Border Bar

This is available below the ribbon. It consists of all the options to filter the objects that can be selected from the graphics window.

Menu

Menu is located on the Top Border Bar. It consists of various options (menu titles). When you click on a menu title, a drop-down appears. You can select the required option from this drop-down.

Status bar

This is available below the graphics window. It displays the prompts and the action taken while using the tools.

Select objects and use MB3, or double-click an object

Resource Bar

This is located at the left side of the window. It contains all the navigator windows such as Assembly Navigator, Constraint Navigator, Part Navigator, and so on.

Part Navigator

Contains the list of operations carried while constructing a part.

Roles Navigator

The Roles Navigator (click the **Roles** tab on the **Resource Bar**) contains a list of system default and industry specific roles. A role is a set of tools and ribbon tabs customized for a specific application. For example, the **CAM Express** role can be used for performing manufacturing operations. This textbook uses the **Advanced** role.

High Definition

The High Definition role displays extra-large icons on the screen. It is suitable for 4K high definition monitors.

The Touch Panel and Touch Tablet roles help you to work with a Multi-touch screen.

Touch Panel

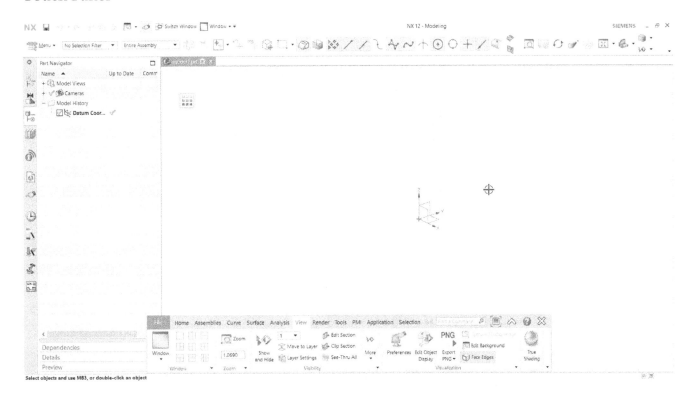

Touch Tablet

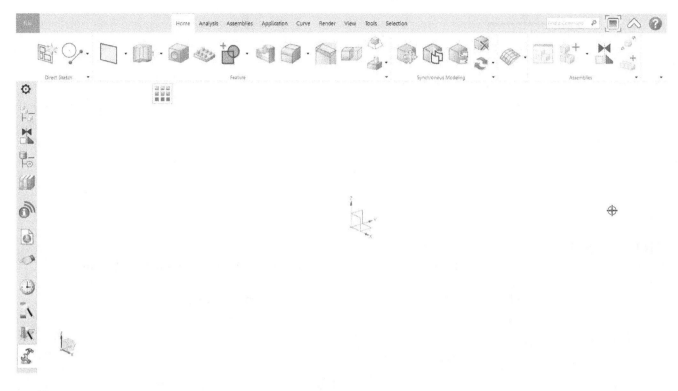

Dialogs

When you execute any command in NX, the dialog related to it appears. The dialog consists of various options. The following figure shows various components of the dialog.

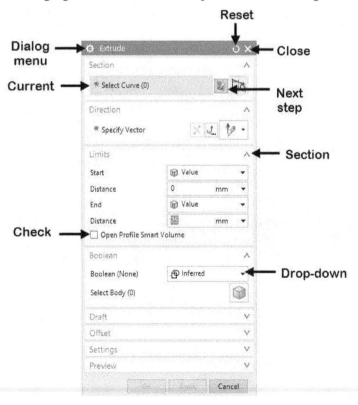

This textbook uses the default options in the dialog. If you have made any changes on the dialog, click the **Reset** button to display the default options.

Mouse Functions

Various functions of the mouse buttons are discussed next.

Left Mouse button (MB1)

When you double-click the left mouse button (MB1) on an object, the dialog related to the object appears. Using this dialog, you can edit the parameters of the objects.

Middle Mouse button (MB2)

Click this button to execute the **OK** command.

Right Mouse button (MB3)

Click this button to display the shortcut menu.

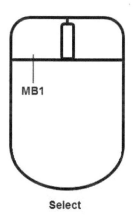
Select

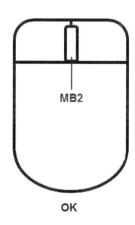
OK

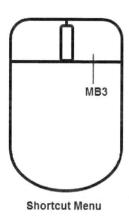
Shortcut Menu

The other functions with combination of the three mouse buttons are given next.

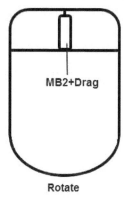
Rotate

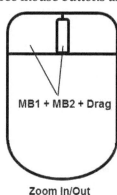
Zoom In/Out

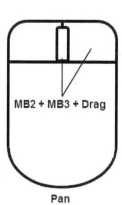
Pan

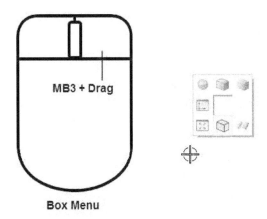

Box Menu

Color Settings

To change the background color of the window, click **View > Visualization > More > Edit Background**; the **Edit Background** dialog appears. Click the **Plain** option to change the background to plain. Click on the color swatches; the **Color** dialog appears. Change the background color and click **OK** twice.

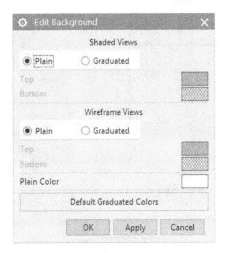

Shortcut Keys

CTRL+Z	(Undo)
CTRL+Y	(Repeat)
CTRL+S	(Save)
F5	(Refresh)
F1	(NX Help)
CTRL+M	(Starts the Modeling environment)
CTRL+SHIFT+D	(Starts the Drafting environment)
CTRL+SHIFT+M	(Starts the NX Sheet Metal environment)
CTRL+ALT+M	(Starts the Manufacturing environment)
X	(Extrude)
CTRL+1	(Customize)
CTRL+D	(Delete)
CTRL+N	(New File)
CTRL+O	(Open File)
CTRL+P	(Plot)

Chapter 2: Modeling Basics

This chapter takes you through the creation of your first NX model. You construct simple parts using NX modeling commands:

In this chapter, you will:

- Construct Sketches
- Construct a base feature
- Add another feature to it
- Construct revolved features
- Apply draft

TUTORIAL 1

This tutorial takes you through the creation of your first NX model. You construct the Disc of an Old ham coupling:

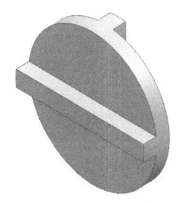

Starting a New Part File

1. To start a new part, click the **New** button on the ribbon; the **New** dialog appears.
2. The **Model** template is the default selection, so click **OK**; a new model window appears.

Starting a Sketch

1. To start a new sketch, click the **Sketch** button on the **Direct Sketch** group; the **Create Sketch** dialog appears.

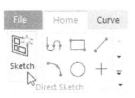

2. Select the XZ plane.

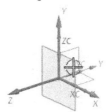

3. Click the **OK** button on **the Create Sketch** dialog; the sketch starts.

The first feature is an extruded feature created from a sketched circular profile. You will begin by sketching the circle.

4. Click **Circle** on the **Direct Sketch** group.
5. Move the pointer to the sketch origin, and then click.
6. Drag the pointer and click to draw a circle.

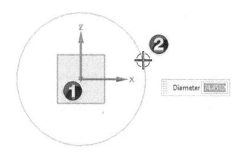

7. Press **ESC** to quit the tool.

Adding Dimensions

In this section, you will specify the size of the sketched circle by adding dimensions.

Note: You may notice that dimensions are applied automatically. However, they do not constraint the sketch. You can hide/show these auto-dimensions using the **Display Sketch Auto Dimensions** icon.

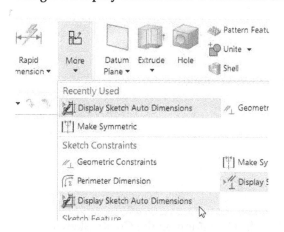

As you add dimensions, the sketch can attain any one of the following three states:

Fully Constrained sketch: In a fully constrained sketch, the positions of all the entities are fully described by dimensions or constraints or both. In a fully constrained sketch, all the entities are dark green color.

Under Constrained sketch: Additional dimensions or constraints or both are needed to completely define the geometry. In this state, you can drag the sketch elements to modify the sketch. An under constrained sketch element is in maroon color.

Over Constrained sketch: In this state, an object has conflicting dimensions or relations or both. An over

constrained sketch entity is grey. The over constraining dimensions are in red color.

1. Double-click on the dimension displayed on the sketch; the **Dimension** edit box appears.
2. To change the dimension to 100 mm, type the value in the **Dimension** edit box, and then press **Enter**.
3. Press **Esc** to quit the **Dimension** tool.

To display the entire circle at a full size and to center it in the graphics area, click **Fit** on **Top Border Bar**.

4. On the ribbon, click **Home > Direct Sketch > Finish Sketch** .
5. To change the view to isometric, click **Orient View Drop-down > Isometric** on the **Top Border Bar**.

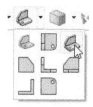

You can use the buttons on the **Orient View** Drop-down on the **Top Border Bar** to set the view orientation of the sketch, part, or assembly.

Constructing the Base Feature

The first feature in any part is called the base feature. In this example, you construct this feature by extruding the sketched circle.

1. On the ribbon, **Home > Feature > Extrude** ; the **Extrude** dialog appears.
2. Click on the sketch.
3. Type-in 10 in the **End** box attached to the preview.
4. To see how the model would look if you have extruded the sketch in the opposite direction, click **Reverse Direction** button in the **Direction** section. Again, click on it to extrude the sketch in the front direction.
5. Ensure that **Body Type** in **Settings** group is set to **Solid**.

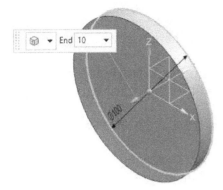

6. Click **OK** to construct the extrusion.

Notice the new feature, **Extrude**, in the **Part Navigator**.

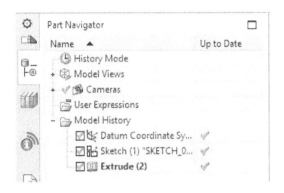

To magnify a model in the graphics area, you can use the tools on the **Zoom** group of the **View** tab of the ribbon.

Click **Zoom**, and then drag the pointer to draw a rectangle; the area in the rectangle zooms to fill the window.

Click **Zoom In/Out**, and then drag the pointer. Dragging up zooms out; dragging down zooms in. Note that this command is available on the **More** gallery.

Type in a value in the **Zoom Scale** box located on the **Zoom** group; the model in the graphics window is zoomed based on the value that you enter.

In addition to the **Zoom** group, the **View** group of the Top Border Bar provides you some more tools.

Click **Fit** to display the part full size in the current window.

Click a vertex, an edge, or a feature, and then click **Fit View to Selection**; the selected item zooms to fill the window.

To display the part in different modes, select the o the **Style** drop-down on the **View** group of the Top Border Bar.

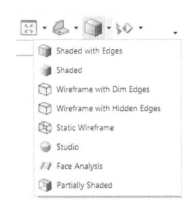

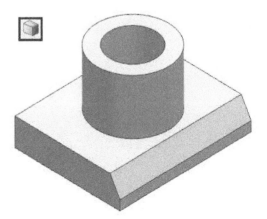

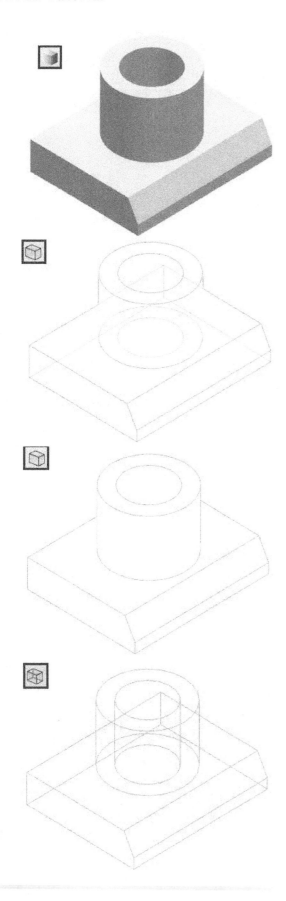

The default display mode for parts and assemblies is **Shaded with Edges**. You may change the display mode whenever you want.

Adding an Extruded Feature

To construct additional features on the part, you need to sketch on the model faces or planes, and then convert them into features.

1. Click **Static Wireframe** ⬦ from the **Style** drop-down on the **View** group of the Top Border Bar.
2. Click **Sketch** on the **Direct Sketch** group of the **Home** ribbon tab.
3. Click on the front face of the part to select it, and then click **OK**.
4. Click **Direct sketch > More Curve > Project Curve** on the ribbon; the **Project Curve** dialog appears.
5. Click on the circular edge.

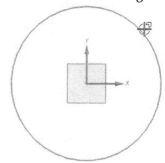

6. Click **OK** on the **Project Curve** dialog; the circular edge projects onto the sketch plane.

7. Click **Line** ✎ on the **Direct Sketch** group.
8. Click on the circle to specify the first point of the line.

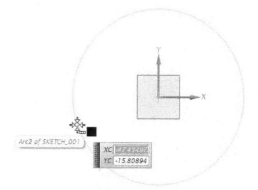

9. Move the pointer towards right.
10. Click on the circle; a line is drawn.

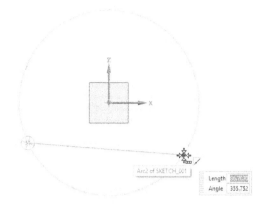

11. Draw another line above the previous line.
12. Press **Esc**.

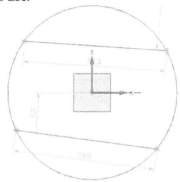

Adding Constraints and Dimensions to the Sketch

To establish the location and size of the sketch, you have to add the necessary constraints and dimensions.

1. Select the lower horizontal line.
2. On the **Shortcuts** toolbar, click **Horizontal** .
3. On the ribbon, click **Direct Sketch > More gallery > Sketch Constrains > Make Symmetric** .
4. Select the first and second lines.
5. Select the X-axis as the centerline; the two lines become symmetric about the X-axis.

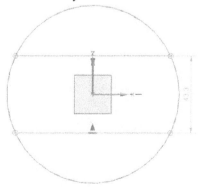

6. Click **Close** on the **Make Symmetric** dialog.

Adding Dimension

1. Double-click on the dimension displayed in the sketch.
2. Type-in 12 in the box displayed.
3. Click **Close** on the dialog.

Trimming Sketch Entities

1. Click **Trim Recipe Curve** on the **Direct Sketch** group.
2. Click on the projected element.
3. Click on the two horizontal lines.
4. On the **Trim Recipe Curve** dialog, click **Discard** under the **Region** section.
5. Click **OK** to trim the projected elements.

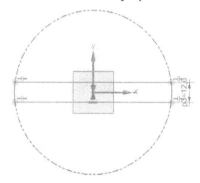

6. Click **Finish Sketch** on the **Direct Sketch** group.
7. To change the view to isometric, click **View > Orientation > Isometric**.

Extruding the Sketch

1. Click on the sketch, and then click **Extrude** on the **Shortcuts toolbar**; the **Extrude** dialog appears.

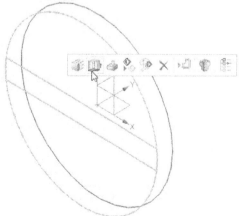

2. Type-in 10 in the **End** box attached to the preview.

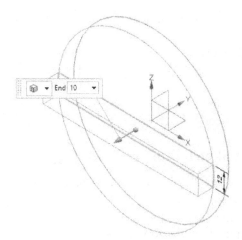

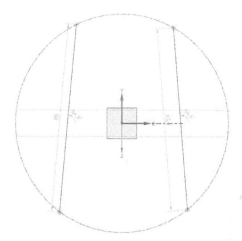

3. Click **OK** to construct the extrusion.
4. To hide the sketch, click **View> Show and Hide**
.
5. On the **Show and Hide** dialog, click **Hide** in the **Sketches** row; the sketches are hidden.
6. Click **Close** on the **Show and Hide** dialog.

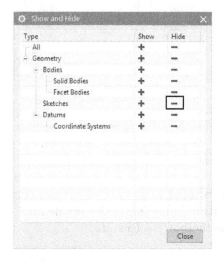

Adding another Extruded Feature

1. Draw a sketch on the back face of the base feature:
   - Use the **Line** command to create the lines, as shown.

- Select a line, and then click the **Vertical** icon on the **Shortcuts** toolbar; the selected line made vertical.
- Likewise, make the other line vertical.
- Make the lines symmetric about the Y axis.
- Add 12 mm dimension between the lines.
- Project the outer circular edge.
- Use the **Trim Recipe Curve** command to trim the projected curve.

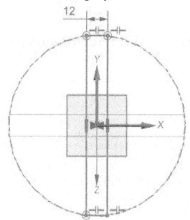

You can use the **Rotate** button from the **View** tab to rotate the model.

2. Click **Finish Sketch** on the **Direct Sketch** group.
3. Extrude the sketch up to 10 mm thickness.

To move the part view, click **View > Orientation > Pan**, then drag the part to move it around in the graphics area.

4. Click **View > Style > Shaded with Edges** on the ribbon.

Replaying the Features of the model

1. On the Top Border Bar, click **Menu > Tools > Update > Feature Replay** .
2. On the **Feature Replay** dialog, expand the Settings section.
3. Type 2 in the **Seconds between Steps** box.
4. Click the **Start** icon in the **Replay Control** section of the **Feature Replay** dialog.
5. Click the **Play** icon to replay the features created in the step-by step order.
6. Click **Close** on the **Replay Feature** dialog.

Saving and closing the Part

7. Create a new folder NX 12 and a sub-folder C2 inside it.
8. Click **Save** on the **Quick Access Toolbar**; the **Name Parts** dialog appears.
9. Type-in **Disc** in the **Name** box and click the **Folder** button.
10. Browse to the NX 12/C2 folder and then click the **OK** button twice.
11. Click **File tab > Close > All Parts** to close the opened file.

Note:
*.prt is the file extension for all the files constructed in the Modeling, Assembly, and Drafting environments of NX.

TUTORIAL 2

In this tutorial, you construct a flange by performing the following actions:

- Constructing a revolved feature
- Constructing a cut features
- Adding fillets

Open a New Part File

1. To open a new part, click **File > New** on the ribbon; the **New** dialog appears.
2. The **Model** is the default selection, so click **OK**; a new model window appears.

Sketching a Revolve Profile

You construct the base feature of the flange by revolving a profile around a centerline.

1. Click the **Sketch** button on the **Direct Sketch** group.
2. Select the YZ plane.
3. Click the **OK** button; the sketch starts.
4. Click **Profile** on the **Direct Sketch** group.
5. Draw a sketch similar to that shown in figure. Press **Esc**.

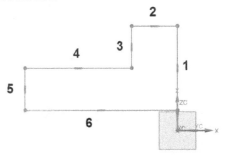

6. On the ribbon, click **Direct Sketch > More gallery > Sketch Constraints > Geometric Constraints** .
7. Click **OK** on the message box.
8. On the **Geometric Constraints** dialog, click **Collinear** .
9. Under the **Geometry to Constrain** section, check **Automatic Selection Progression**.
10. Click on the line 1 and the Y-axis to make them collinear.

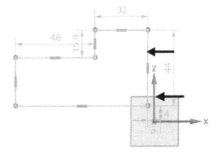

11. Click **Rapid Dimension** on the **Direct Sketch** group.

12. Select the X-axis and Line 6; a dimension appears.
13. Place the dimension and type-in 15 in the dimension box.
14. Press **Enter** key.

Note: The first dimension of a sketch defines its size. It makes the sketch larger or smaller.

15. Select the X-axis and Line 4; a dimension appears.
16. Set the dimension to 30.
17. Select the X-axis and Line 2; a dimension appears.
18. Set the dimension to 50 mm.
19. Create a dimension between the Y-axis and Line 3.
20. Set the dimension to 20 mm.
21. Create a dimension of 50 mm between Y-axis and Line 5.
22. Close the **Rapid Dimension** dialog.

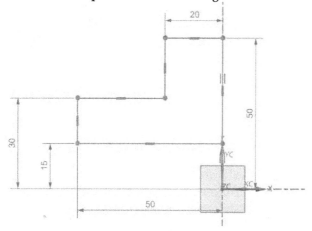

23. Onn the ribbon, click **Home > Direction Sketch > More > Sketch Tools > Relations Browser**.

24. On the **Sketch Relations Browser** dialog, select **Scope > All in Active Sketch**.
25. Select **Constraints** from the **Top-level Node Objects** section.

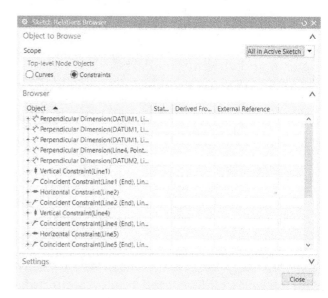

Notice the constraints and dimensions created inside the sketch.

26. In the **Sketch Relations Browser** dialog, right click on a constraint and select **Fit View to Selection**; the selected constraint is zoomed.

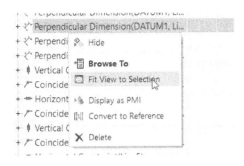

You can also hide, delete or convert a dimension into reference.

27. Close the **Sketch Relations Browser** dialog.
28. Click **Finish Sketch** on the **Direct Sketch** group.
29. To change the view to isometric, click **Isometric** on the **View** tab.

Constructing the Revolved Feature

1. On the ribbon, click **Home > Feature > Extrude > Revolve**; the **Revolve** dialog appears.

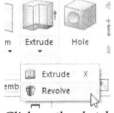

2. Click on the sketch.
3. Click on **Specify Vector** in the **Axis** group; a vector triad appears.
4. Click on the Y-axis of the triad.

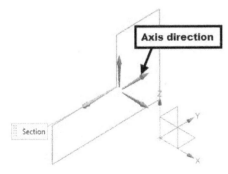

5. Click on the origin point of the Coordinate system; the preview of the revolved feature appears.

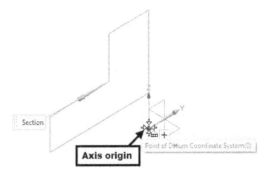

6. Type-in 360 in the **End** box attached to the preview.
7. Click **OK** to construct the revolved feature.

Constructing the Cut feature

1. Click **Extrude** on the **Feature** group.
2. Rotate the model geometry and click the back face of the part; the sketch starts.
3. On the **Home** tab of the ribbon, click the **Curve** drop-down on the **Curve** group.

4. Select the **Project Curve** icon from the **Curve** gallery.

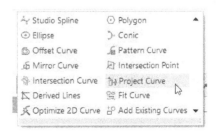

5. Click on the outer circular edge of the model.

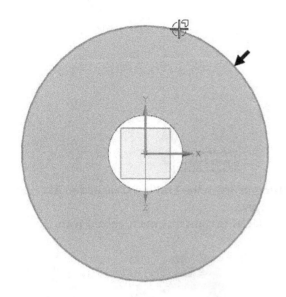

6. Click **OK** on the **Project Curve** dialog.
7. On the ribbon, click **Home** tab > **Curve** group > **Rectangle** 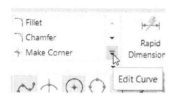.
8. On the **Rectangle** dialog, click the **From Center** icon.
9. Select the sketch origin.
10. Move the pointer horizontally toward right, and then click.

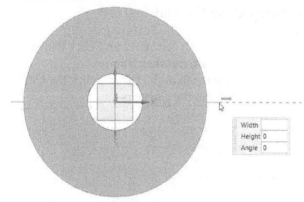

11. Move the pointer upward and click.

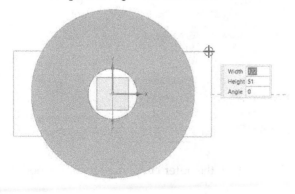

12. On the **Curve** group of the ribbon, click the **Edit Curve** drop-down.

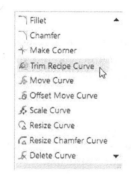

13. Click the **Trim Recipe Curve** icon.

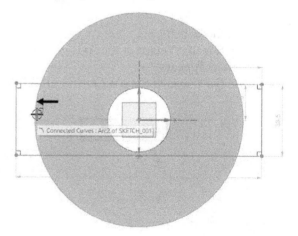

14. On the **Trim Curve** dialog, select the **Discard** option from the **Region** section.
15. Click on the portion of the circular edge enclosed by the rectangle.

16. Click on the horizontal line of the rectangle.
17. Click **OK** on the **Trim Recipe Curve** dialog.

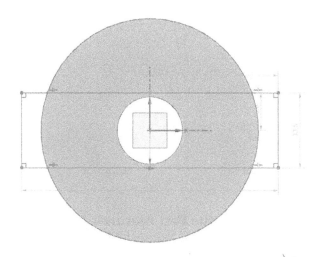

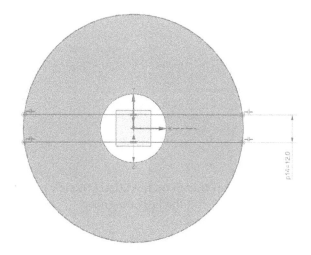

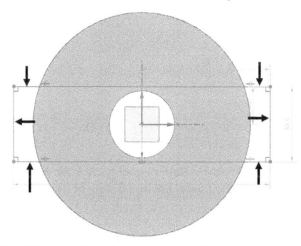

18. On the **Curve** group, click the **Quick Trim** icon.
19. Click on the lines outside the model, as shown.

20. Click the **Close** button on the dialog.

21. Use the **Make Symmetric** command to make the horizontal lines symmetric about the X-axis.

22. Use the **Rapid Dimension** command to add the dimension of 12 mm between the two horizontal lines.

23. Click **Finish** on the **Sketch** group.
24. Enter 10 in the **End** box attached to the preview.
25. Click **Reverse Direction** in the **Direction** section.
26. Select **Subtract** in the **Boolean** section.

27. Click **OK** to construct the cut feature.

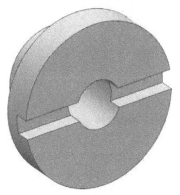

Adding another Cut-out

1. Draw a sketch on the front face of the model geometry (Use the **Profile** command to create the lines, and then project the inner circular edge. Use the **Trim Recipe Curve** command to trim the projected curve. Also, add dimensions).

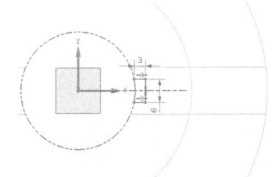

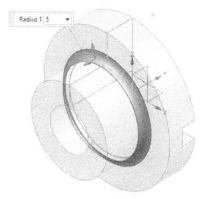

2. Click **Finish sketch** on the **Direct Sketch** group.
3. Click **Extrude** on the **Feature** group.
4. Click on the sketch.
5. On the **Extrude** dialog, select **End > Through All** under the **Limits** section.
6. Click **Reverse Direction** in the **Direction** section.
7. Select **Subtract** in the **Boolean** group.
8. Click **OK** to construct the cut-out feature.
9. To change the view to isometric, click **Isometric** on the **View** tab.

Adding Edge blends

1. Click **Home > Feature > Edge Blend** ; the **Edge Blend** dialog appears.
2. Click on the inner circular edge and set **Radius 1** to 5.
3. Click **OK** to add the blend.

Playing back the Part Features

1. On the Top Border Bar, click **Menu > Edit > Feature > Replay** ; the **Feature Replay** dialog appears.

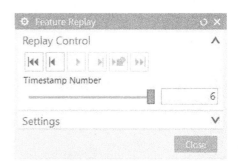

2. On the **Feature Replay** dialog, click the **Start** button; the model the initial state of the model is displayed.
3. Expand the **Settings** section of the **Feature Replay** dialog and enter 1 in the **Seconds between Steps** box.
4. Click the **Play** button on the **Feature Replay** dialog; NX plays the sequence in which the model was created.
5. Close the **Feature Replay** dialog.

Saving and Closing the Part

1. Click **File > Save > Save**; the **Name Parts** dialog appears.
2. Type **Flange** in the **Name** box, and then click the **Folder** button next to it.
3. Browse to NX 12/C2 folder and then click **OK** button twice.
4. Click **File > Close > All Parts**.

TUTORIAL 3

In this tutorial, you construct a Shaft by performing the following:

- Constructing a revolved feature
- Constructing a cut feature

Opening a New Part File

1. To open a new part, click the **New** button on the **Standard** group.

2. Select the **Model** template and click **OK**; a new model window appears.

Constructing the Revolved Feature

1. Click **Extrude > Revolve** on the **Feature** group.

2. Click the **Sketch Selection** icon under the **Section** section of the **Revolve** dialog.

3. Click on the YZ plane to select it, and then click **OK**; the sketch starts.

4. On the ribbon, click **Home > Curve > Rectangle** .

5. On the **Rectangle** dialog, click the **By 2 Points** icon.

6. Select the origin point of the sketch.

7. Move the pointer toward top left corner and click.

8. Add dimensions to the sketch, as shown in figure.

9. Click **Finish** on the **Sketch** group.

10. Click on the Y-axis of the triad.

11. Click on the origin point of the coordinate system; the preview appears.

12. Click **OK** to construct the revolved feature.

Creating Cut feature

1. Click **Home > Direct Sketch > Sketch** on the ribbon.

2. Click on the front face of the model, and then click **OK**.

3. On the ribbon, click **Home > Direct Sketch > Rectangle**.

4. On the **Rectangle** dialog, click the **By 2 Points** icon.

5. Create a rectangle, as shown.

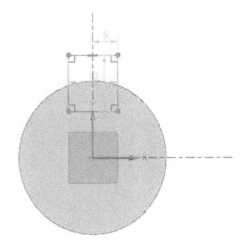

6. On the ribbon, click **Home > Direct Sketch > Sketch Curve** drop-down **> More Curve > Project Curve** .

7. Click on the circular edge of the model.

8. Click **OK**.

9. On the ribbon, click **Home > Direct Sketch > Trim Recipe Curve** .

10. Click on the portion of the projected curve outside the rectangle.

11. Select the rectangle to define the boundary edges.

12. Select the **Discard** option from the **Region** section.

13. Click **OK** to trim the selected portion of the projected curve.

14. Activate the **Quick Trim** command, and then select the entities of the rectangle outside the model.

15. Click **Close** on the **Quick Trim** dialog.

16. Use the **Make Symmetric** command to make the vertical lines of the sketch symmetric about the Y-axis.

17. Add dimensions to the sketch using the Rapid Dimension command.

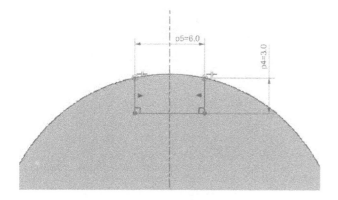

18. Finish the sketch.
19. Click **Extrude** on the **Feature** group.
20. Click on the sketch.
21. Type-in 55 in the **End** box.
22. Click **Reverse Direction** in the **Direction** section.
23. Select **Subtract** from the **Boolean** group.

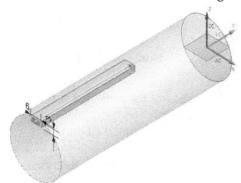

24. Click **OK** to construct the cut feature.

Saving the Part
1. Click **File > Save > Save**; the **Name Parts** dialog appears.
2. Type **Shaft** in the **Name** box and click **Folder** button.
3. Browse to NX 12/C2 folder and then click **OK** button twice.
4. Click **File > Close > All Parts**.

TUTORIAL 4
In this tutorial, you construct a Key by performing the following:

- Constructing a Block
- Applying draft

Constructing Extruded feature
1. Open a new part file.
2. On the ribbon, click **Home > Feature > More >**

   Design Feature > Block 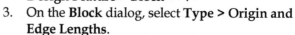.
3. On the **Block** dialog, select **Type > Origin and Edge Lengths**.
4. Type-in **6**, **50**, and **6** in the **Length (XC)**, **Width (YC)**, and **Height (ZC)** boxes, respectively.
5. Click on the origin point of the datum coordinate system.

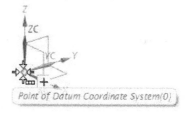

6. Click **OK** to construct the block.

Applying Draft
1. Click **Draft**  on the **Feature** group.
2. On the **Draft** dialog, select **Type > Face**.
3. Click on Y-axis to specify vector.

4. Select front face as the stationary face.

5. Click **Select Face** in the **Faces to Draft** section.
6. Select the top face.

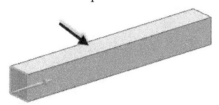

7. Type-in 1 in the **Angle 1** box.
8. Click **OK** to add the draft.

Saving the Part

1. Click **File > Save > Save**; the **Name Parts** dialog appears.
2. Type **Key** in the **Name** box and click **Folder** button.
3. Browse to NX 12/C2 folder and then click **OK** button twice.
4. Click **File > Close > All Parts**.

Chapter 3: Constructing Assembly

In this chapter, you will:

- Add Components to an assembly
- Apply constraints between components
- Produce exploded view of the assembly

TUTORIAL 1

This tutorial takes you through the creation of your first assembly. You construct the Oldham coupling assembly:

Copying the Part files into a new folder

1. Create a folder named **Oldham_Coupling** at the location NX 12/C3.
2. Copy all the part files constructed in the previous chapter to this folder.

Opening a New Assembly File

1. To open a new assembly, click **File > New**; the **New** dialog appears.
2. Click **Assembly** in the **Templates** group.
3. Click **OK**; a new assembly window appears. In addition, the **Add Component** dialog appears.

Inserting the Base Component

1. To insert the base component, click **Open** button in the **Part To Place** section of the **Add Component** dialog.
2. Browse to the location NX 12/C3/Oldham_Coupling and double-click on **Flange.prt**.
3. On the **Add Component** dialog, select **Component Anchor > Absolute** in the **Location** section.
4. Select **Assembly Location > WCS**.

5. Under the **Settings** section, select **Reference Set > Entire Part**. The **Entire Part** option displays all the datum planes, sketches, and features of the part.
6. Click **OK** to place the Flange at the origin.

The **Create Fix Constraint** message box appears on the screen.

7. Click **No** on the **Create Fix Constraint** message box.

There are two ways of constructing any assembly model.

- Top-Down Approach
- Bottom-Up Approach

Top-Down Approach
You open the assembly file, and then construct components files in it.

Bottom-Up Approach
You construct the components first, and then add them to the assembly file. In this tutorial, you construct the assembly using this approach.

Adding the second component

1. To insert the second component, click

   Assemblies > Component > Add on the

ribbon; the **Add Component** dialog appears.

2. On the **Add Component** dialog, click **Open** button in the **Part To Place** section.

3. Browse to the location NX 12/C3/Oldham_Coupling and double-click on **Shaft.prt**.

4. Under the **Placement** section, select the **Constrain** option.

5. Under the **Settings** section, select **Reference Set > Entire Part**.

After adding the components to the assembly environment, you have to apply constraints between them. By applying constraints, you establish relationships between components. You can apply the following types of constraints between components.

Touch Align: Using this constraint, you can make two faces coplanar to each other. Note that if you set the **Orientation** to **Align**, the faces will point in the same direction. You can also align the centerlines of the cylindrical faces.

Concentric: This constraint makes the centers of circular edges coincident. In addition, the circular edges will be on the same plane.

Distance: This constraint provides an offset distance between two objects.

Fix: This constraint fixes a component at its current position.

Parallel: This constraint makes two objects parallel to each other.

Perpendicular: This constraint makes two objects perpendicular to each other.

Fit: This constraint brings two cylindrical faces together. Note that they should have the same radius.

Bond: This constraint makes the selected components rigid so that they move together.

Center: This constraint positions the selected component at a center plane between two components.

Angle: Applies angle between two components.

Align/Lock: Aligns the axes of two cylindrical faces and locks the rotation.

6. On the **Add Components** dialog, select **Constraint Type > Touch Align**.

7. Under the **Geometry to Constrain** section, select **Orientation > Infer Center/Axis**.

8. Under the **Settings** section, uncheck the **Preview** option.

9. Check the **Enable Preview Window** option; the **Component Preview** window appears.

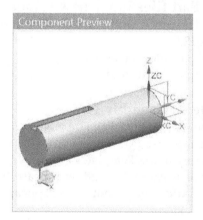

10. Click on the cylindrical face of the Shaft.

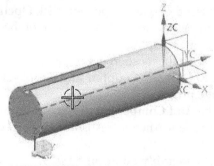

11. Click on any cylindrical face of the Flange.

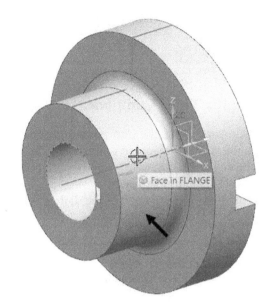

12. Under the **Geometry to Constrain** section, select **Orientation > Align** .

13. Click on the front face of the shaft.

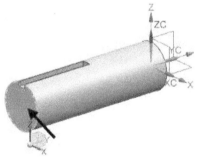

14. Rotate the flange and click on the slot face as shown in figure.

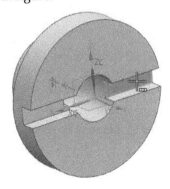

15. Click on the YZ plane of the Shaft.

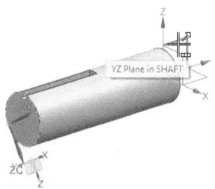

16. Click on the XY plane of the Flange.

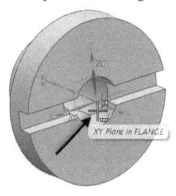

17. Click **OK** to assemble the components.

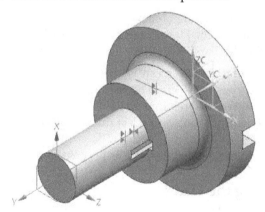

Checking the Degrees of the Freedom

1. To check the degrees of freedom of a component, click **Assemblies > Component Position > Show Degrees of Freedom** .
2. Click on the Flange to display the degrees of freedom.

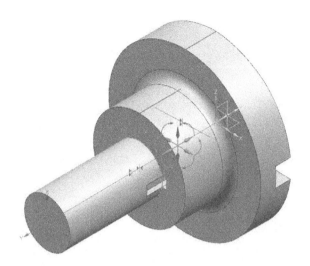

You will notice that the Flange has six degrees of freedom.

Fixing the Flange

1. To fix the flange, click **Assemblies > Component Position > Assembly Constraints** on the ribbon.
2. On the **Assembly Constraints** dialog, click **Constraint Type > Fix**.
3. Click on the Flange, and then click **OK**.
4. On the Top Border Bar, click **Menu > View > Operation > Refresh**.
5. To view the degrees of freedom, click **Show Degrees of Freedom** on the **Component Position** group and select the Flange and Shaft.

You will notice that they are fully constrained.

Hiding the Flange

1. To hide the Flange, click on it and select **Hide** from the contextual toolbar.

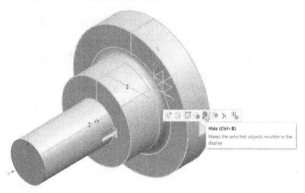

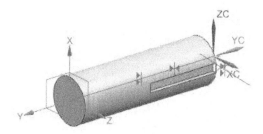

Adding the Third Component

1. Click **Add** on the **Component** group.
2. On the **Add Component** dialog, click the **Open** button.
3. Double-click on the **Key.prt**.
4. In the **Placement** section of the dialog, select **Constraint Type > Touch Align**.
5. Under the **Geometry to Constraints** section, select **Orientation > Align**.
6. Click **Select Two Objects** in the **Geometry to Constrain** section.
7. Click on the front face of the Key and front face of the Shaft.

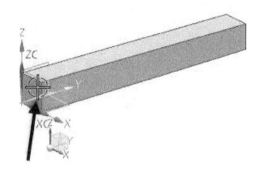

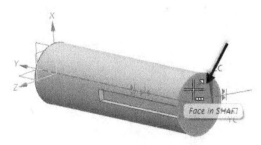

8. Click on the XY plane of the Key.

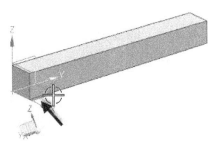

9. Click on the face on the shaft, as shown in figure.

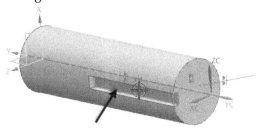

10. Under the **Geometry to Constrain** section, select **Orientation > Touch.**
11. Click on the side face of the Key and select the face on shaft as shown in figure.

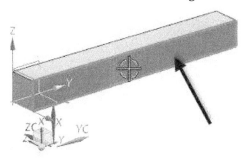

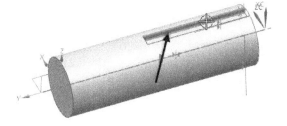

8. Click **OK.**

Showing the Hidden Flange

1. To show the hidden flange, click the **Assembly Navigator** tab, right click on the Flange, and then select **Show** .

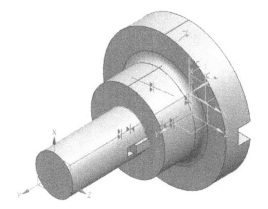

Hiding the Reference Planes, sketches, and Constraint symbols

1. To hide the reference planes, sketches, and constraint symbols, click **View > Visibility > Show and Hide** on the ribbon.
2. On the **Show and Hide** dialog, click the hide icons in the **Sketches**, **Datums**, and **Assembly Constraints** rows.

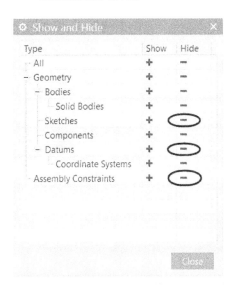

3. Click **Close** on the dialog.

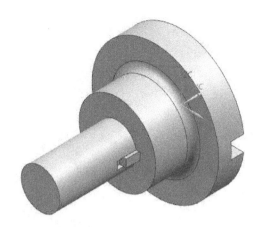

Saving the Assembly

1. Click **File > Save > Save**; the **Name Parts** dialog appears.
2. Type-in **Flange_subassembly** in the **Name** box and click the **Folder** button.
3. Browse to NX 12/C3/Oldham_Coupling folder and then click the **OK** button twice.
4. Click **File > Close > All Parts**.

Starting the Main assembly

1. Click **File > New** on the ribbon.
2. On the **New** dialog, click the **Assembly** template.
3. Type-in **Main_assembly** in the **Name** box and click **Folder** button.
4. Browse to NX 12/C3/ Oldham_Coupling folder and then click **OK** button twice; the **Add Component** dialog appears.

Adding Disc to the Assembly

1. Click the **Open** button.
2. Double-click on **Disc.prt**.
3. Under the **Location** section, select **Component Anchor > Absolute**.
4. Select **Assembly Location > WCS**.
5. Under the **Placement** section, select the **Constrain** option.
6. Select **Constraint Type > Fix**.
7. Set **Reference Set** to **Model**.
8. Select the Disc from the **Component Preview** window.
9. Click **OK** to place the Disc at the origin.

Placing the Sub-assembly

1. Click the **Add** button on the **Component** group.
2. Click the **Open** button.

3. Double-click on **Flange_subassembly.prt**.
4. On the **Add Component** dialog, select **Placement > Constrain**.
5. Set **Constraint Type** to **Touch Align**.
6. Set **Orientation** to **Touch**.
7. Click **Select Two Objects** from the **Geometry to Constrain** section.
8. Click on the face of the Flange as shown in figure.

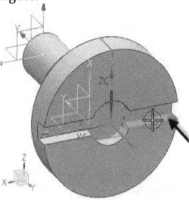

9. Click on the face of the Disc as shown in figure.

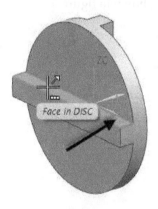

10. Set **Constraint Type** to **Concentric**.
11. Click on the circular edge of the Flange.

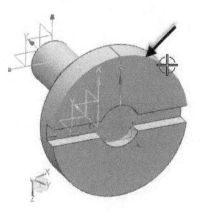

12. Click on the circular edge of the Disc.

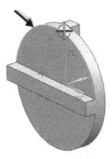

13. Click **OK** to assemble the subassembly.

Placing the second instance of the Sub-assembly

1. Insert another instance of the Flange subassembly.
2. Apply the **Touch Align** and **Concentric** constraints. Note that you have to click the **Reverse Last Constraint** button while applying the **Concentric** constraint.

Saving the Assembly

1. Click **Save** on the **Quick Access Toolbar**, or click **File > Save**.

TUTORIAL 2

In this tutorial, you produce the exploded view of the assembly created in the previous tutorial.

Producing the Exploded view

1. To produce the exploded view, click **Exploded Views > New Explosion**; the **New Explosion** dialog appears.
2. Type-in Oldham_Explosion in the **Name** box.
3. Click **OK**.
4. On the **Assembly Navigator**, click the right mouse button on Flange_subassembly x 2.

5. Select **Unpack**.
6. Press Esc to deselect the **Flange_subassembly** instances.
7. Click **Exploded Views > Edit Explosion** on the ribbon; the **Edit Explosion** dialog appears.
8. Select Flange_subassembly from the **Assembly Navigator**.

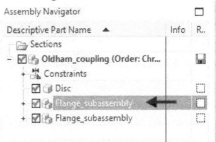

9. Click **Move Objects** on the dialog; the dynamic triad appears on the flange subassembly.

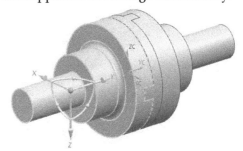

10. Click the **Snap Handles to WCS** button on the **Edit Explosion** dialog; the dynamic triad snaps to WCS.

11. Click the **Y-Handle** on the dynamic triad.
12. Enter **-100** in the **Distance** box.
13. Click **OK** to explode the flange subassembly.

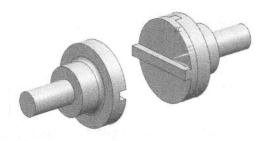

14. Click **Edit Explosion** button on the **Exploded Views** group.
15. Click **Select Objects** on the **Edit Explosion** dialog.
16. Rotate the model and select the Key from the assembly.
17. Click **Move Objects** on the dialog.
18. Click **Snap Handles to WCS** button.

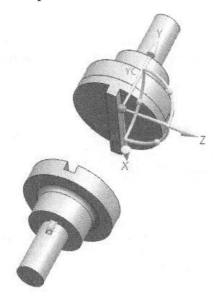

19. Click the **Y-Handle** on the dynamic triad.

20. Enter **80** in the **Distance** box.
21. Click **OK** to explode the Key.

22. Activate the **Edit Explosion** dialog.
23. Explode the shaft in Y-direction up to the distance of **-80 mm**.

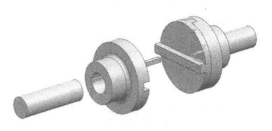

24. Likewise, explode the other flange subassembly and its parts in the opposite direction. The explosion distances are same.

Creating Tracelines

1. To create tracelines, click **Exploded Views > Tracelines** on the ribbon; the **Tracelines** dialog appears.
2. Click on the center point of the Flange.

3. On the **Tracelines** dialog, select **End Object > Point**.
4. Click on the center point of the circular edge of the shaft.

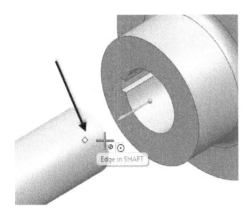

5. Click **OK** to create the traceline.

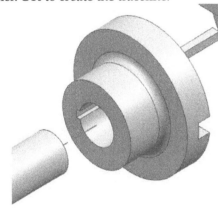

6. Click the **Tracelines** button on the **Exploded Views** group.
7. Under the **Start** section, select **Inferred > End Point** .
8. Select the edge on the key way of the shaft.

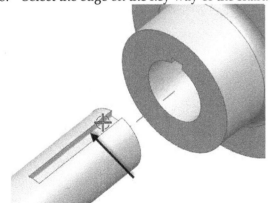

9. Double-click on the arrow displayed on the edge to reverse the direction.
10. Under the **End** section, select **Inferred > End Point**.
11. Click on the edge on the key.

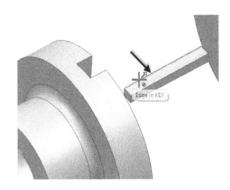

12. Double-click on the arrow to reverse the direction.

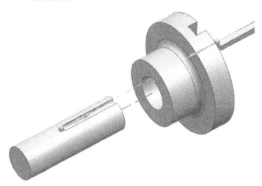

13. Click **OK** to create the traceline.
14. Create tracelines between the other parts.
15. Change the view to **Wireframe with Hidden Edges**.

16. Click **Save** on the **Quick Access Toolbar**, or click **File > Save**.
17. Close the assembly.

Chapter 4: Generating Drawings

In this chapter, you generate drawings of the parts and assembly from previous chapters.

In this chapter, you will:

- Open and edit a drawing template
- Insert standard views of a part model
- Add model and reference annotations
- Add another drawing sheet
- Insert exploded view of the assembly
- Insert the bill of materials of the assembly
- Apply balloons to the assembly

TUTORIAL 1

In this tutorial, you will generate drawings of parts constructed in previous chapters.

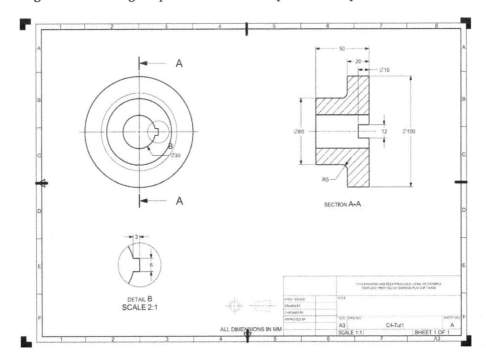

Opening a New Drawing File

1. Start NX 12.
2. To open a new drawing, click the **New** button on the **Standard** group, or click **File > New**.
3. On the **New** dialog, select the **Drawing** tab.
4. Click **A3-Size** in the **Templates** section.
5. Click **OK**; a new drawing window appears. In addition, the **Populate Title Block** dialog appears.
6. Select the individual labels and type-in their values.
7. Click the **Close** button on this dialog.

Editing the Drawing Sheet

1. To edit the drawing sheet, click **Home > New Sheet > Edit Sheet** on the ribbon; the **Sheet** dialog appears.

2. Click the gear icon located at the top left corner of the dialog, and select **Sheet (More)**.

3. Expand the **Settings** section and set **Units** to **Millimeters**.

4. Set the **Projection** type to **3ʳᵈ Angle Projection** ◎ ⊟ .

5. Click **OK** on the **Sheet** dialog.

Generating the Base View

1. To generate the base view, click **Base View** on the **View** group; the **Base View** message box appears.

2. Click **Yes** on the message box; the **Part Name** dialog appears.

3. On the **Part Name** dialog, browse to the location NX 12/C3/Oldham_Coupling and double-click on **Flange.prt**; the **Base View** dialog appears.

   In addition, the view appears along with the pointer.

4. Under the **Model View** section, select **Model View to Use > Front**.

5. Place the view as shown in figure; the **Projected View** dialog appears.

6. Click **Close** to close the dialog.

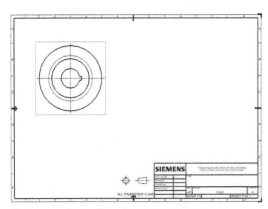

Generating the Section View

1. To generate a section view, click **Home > View > Section View** on the ribbon; the **Section View** dialog appears.

2. Click on the base view; the section line appears.

3. Click on the center point of the base view.

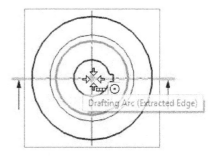

4. Drag the pointer toward right and click to position the section view.

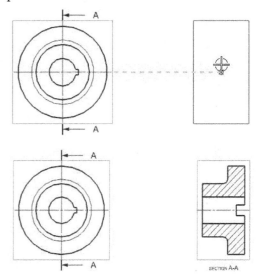

5. Click **Close** on the **Section View** dialog.

Generating the Detailed View

Now, you need to generate the detailed view of the keyway that appears on the front view.

1. To generate the detailed view, click **Detail View** button on the **View** group.

2. On the **Detail View** dialog, select **Type > Circular**.

3. Specify the center point and boundary point of the detail view as shown.

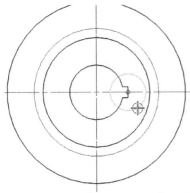

4. Under the **Scale** section, select **Scale > 2:1**.
5. Position the detail view below the base view.
6. Close the **Detail View** dialog.

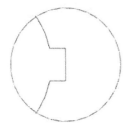

DETAIL **B**
SCALE 2:1

Setting the Annotation Preferences

1. To set the annotation preferences, click **File > Preferences > Drafting**; the **Drafting Preferences** dialog appears.

2. On the dialog, type-in **Orientation and Location** in the **Find** box and press Enter.
3. Set the **Orientation** value to **Horizontal text**.

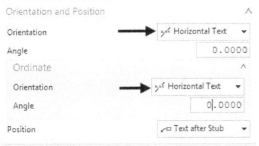

4. On the dialog, select **Dimension > Text > Units** from the tree.
5. Set the **Decimal Delimiter** value to **. Period**.

6. On the dialog, select **Dimension > Text > Dimension Text** from the tree.
7. Under the **Format** section, type-in 3.5 in the **Height** box.
8. Select **Common > Line/Arrow > Arrowhead** from the tree.
9. Under the **Workflow** section, check the **Automatic Orientation** option.
10. Under the **Format** section, type-in **3.5** and **30** in the **Length** and **Angle** boxes, respectively.
11. Click **Common > Line/Arrow > Extension Line**.
12. Type-in **1** in the **Gap** boxes.
13. Type in **2** in the **Extension Line Overhang**.
14. Click **Common > Lettering**.
15. Under the **Text Parameters** section, type-in **3.5** in **Height** box,
16. Click **OK**.

Dimensioning the Drawing Views

1. To add dimensions, click **Home > Dimension > Rapid Dimension** on the ribbon.
2. On the section view, click the horizontal line located at the top.

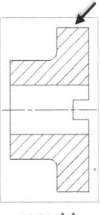

SECTION A-A

3. Drag the pointer up and click to position the dimension.

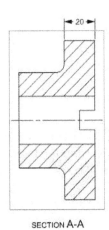

SECTION A-A

4. Click on the ends of the section view, as shown.

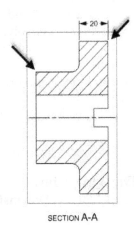

SECTION A-A

5. Drag the pointer up and click to position the dimension.

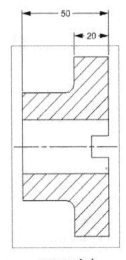

SECTION A-A

6. Add another linear dimension to the section view.

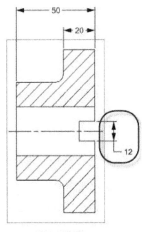

SECTION A-A

7. On the section view, click on the arc located at the bottom.

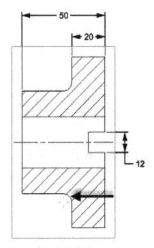

SECTION A-A

8. Drag the pointer downward and click to position the radial dimension.

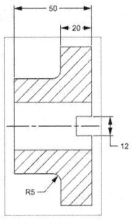

SECTION A-A

9. On the **Rapid Dimension** dialog, under the **Measurement** section, select **Method > Cylindrical**.

10. Click on the ends of the section view, as shown.

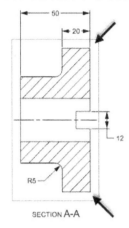

11. Drag the pointer rightwards and click to position the dimension.

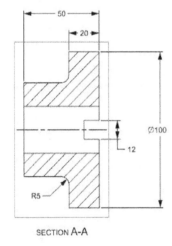

12. Create the other dimensions on the section view, as shown.

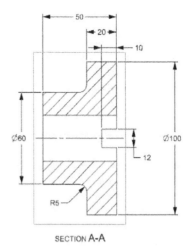

SECTION A-A

13. Create the radial dimension on the front view, as shown.

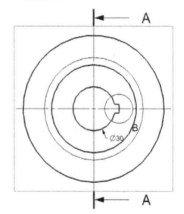

14. Create the dimensions on the detail view, as shown.

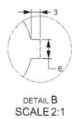

DETAIL B
SCALE 2:1

Saving the Drawing

1. On the **Quick Access Toolbar**, click **Save**; the **Name Parts** dialog appears.
2. Type-in **Flange Drawing** in the **Name** box and click **Folder** button.
3. Browse to NX 12/C3 folder and then click **OK** button twice
4. Close the drawing.

TUTORIAL 2

In this tutorial, you generate the drawing of the Disc constructed in Chapter 1.

Creating a custom template

1. Close the NX 12 application window.
2. Type NX 12 in the search bar located on the Taskbar.
3. Right click on the **NX 12.0** icon and select **Run as administrator**.

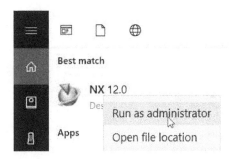

4. Click **Yes** on the message box.
5. On the ribbon, click the **New** button.
6. On the **New** dialog, click the **Model** tab.
7. Double-click on the **Model** template.
8. On the ribbon, click **Application > Design > Drafting** .
9. On the **Sheet** dialog, select **Standard Size**.
10. Set **Size** to **A3 – 297 x 420**.
11. Set **Scale** to **1:1**.
12. Under the **Settings** section, set **Units** to **Millimeters**.
13. Select **3rd Angle Projection** and uncheck **Always Start Drawing View Command**.
14. Click **OK** to open a blank sheet.

Adding Borders and Title Block

1. On the ribbon, click **Drafting Tools > Drawing Format > Borders and Zones** .
2. On the **Borders and Zones** dialog, leave the default settings and click **OK**.

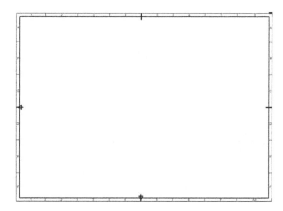

3. On the ribbon, click **Home > Table > Tabular Note** .
4. On the **Tabular Note** dialog, under the **Origin** section, expand the **Alignment** section and select **Anchor > Bottom Right**.
5. Under the **Table Size** section, set **Number of Columns** to **3** and **Number of Rows** to **2**.
6. Type-in **50** in **Column Width** box.
7. Click on the bottom right corner of the sheet border.

8. Click **Close** on the **Tabular Note** dialog.
9. Click on the left vertical line of the tabular note.

10. Press the left mouse button and drag toward right.
11. Release the left mouse button when column width is changed to 35.

12. Likewise, change the width of the second and third columns.

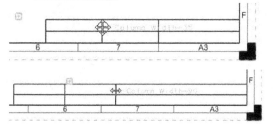

13. Click inside the second cell of the top row.

14. Press the left mouse button and drag to the third cell.

15. Click the right mouse button in the selected cells and select **Merge Cells**.

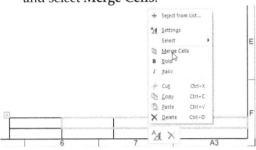

16. Change the height of the top row to 20.

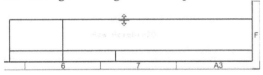

17. Click **Yes** on the message box.
18. Click the right mouse button in the second cell of the top row. Select **Settings** .
19. On the **Settings** dialog, select **Prefix/Suffix** from the tree.
20. Type-in **Title:** in the **Prefix** box.
21. Click **Close**.

22. Likewise, add prefixes to other cells.

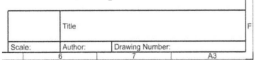

23. Click the right mouse button in the first cell of the top row.
24. Click **Import > Image**.
25. Select your company logo image and click **OK**. Make sure that the size of the image is less than the cell size.
26. On the ribbon, click **Drafting Tools > Drawing**

Format > Define Title Block .

27. Click on the table, and then click **OK**.
28. On the ribbon, click **Drafting Tools > Drawing**

Format > Mark as Template .

29. On the dialog, select **Mark as Template and Update PAX File**.
30. Under the **PAX File Settings** section, type-in **Custom Template** in the **Presentation Name** box.
31. Select **Template Type > Reference Existing Part**.
32. Click the **Browse** icon.
33. Go to
    C:\Program Files\Siemens\NX 12.0\LOCALIZATION\prc\english\startup
34. Click **ugs_drawing_templates**.
35. Click **OK**.
36. On the **Input Validation** box, click **Yes**.
37. Click **OK** twice.
38. Save and close the file.

Opening a new drawing file using the custom template

1. On the ribbon, click the **New** button on the **Home** tab of the ribbon.
2. On the **New** dialog, under the **Drawing** tab, select **Relationship > Reference Existing Part**.
3. Under the **Templates** section, select **Custom Template**.
4. Under the **Part to create a drawing of** section, click the **Browse** button.
5. On the **Select master part** dialog, click **Open** .
6. Go to the location of Disc.prt and double-click on it.
7. Click **OK** twice.
8. On the **Populate Title Block** dialog, type-in values, as shown.
9. Click **Close**.

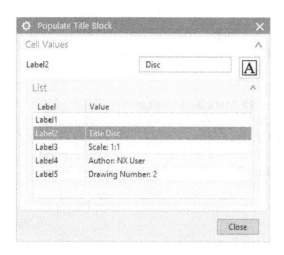

Generating Drawing Views

1. On the **View Creation Wizard** dialog, select **Loaded Parts > Disc.prt**.
2. Click **Next**.
3. On the **Options** page, select **View Boundary > Manual**.
4. Uncheck the **Auto-Scale to Fit** option.
5. Select **Scale > 1:1**.
6. Select **Hidden Lines > Dashed**.

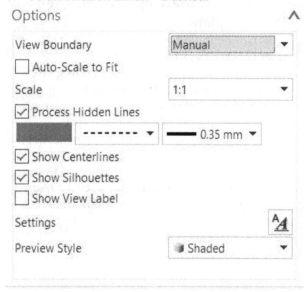

7. Click **Next**.
8. On the **Orientation** page, select **Model Views > Front**.
9. Click **Next**.
10. On the **Layout** page, select the view, as shown.

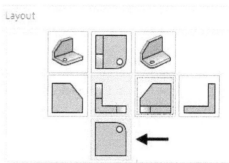

11. Select **Option > Manual**.

12. Click to define the center of the views, as shown.

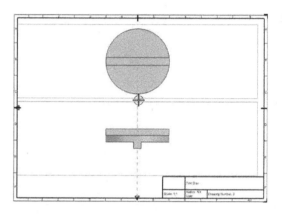

Adding Dimensions

1. Add centerlines and dimensions to the drawing.
2. Save and close the drawing file.

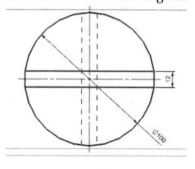

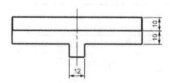

TUTORIAL 3

In this tutorial, you will generate the drawing of Oldham coupling assembly created in the previous chapter.

Creating the assembly drawing

1. Open the Main_assembly.prt file.
2. Click **Applications > Design > Drafting**.
3. On the **Sheet** dialog, select **Standard Size**.
4. Set **Size** to **A3 -297 x 420**.
5. Set **Scale** to **1:2**.
6. Under the **Settings** section, check **Always Start View Creation**.
7. Select **Base View command**.
8. Click **OK**.
9. On the **Base View** dialog, under the **Model View** section, select **Model View to Use > Isometric**.
10. Under the **Scale** section, select **Scale > 1:2**.
11. Click on the left side of the drawing sheet.
12. Click **Close** on the **Projected View** dialog.

Generating the Exploded View

1. On the ribbon, click **Home > View > Base View**.
2. On the **Base View** dialog, select **Model View to Use >Trimetric**.
3. Click on the right side of the drawing sheet.
4. Click **Close**.

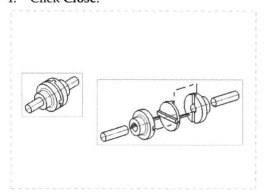

Generating the Part list

1. To generate a part list, click **Home > Table > Part List** on the ribbon.

2. Place the part list at the top-right corner.

Generating Balloons

1. To generate balloons, click **Home > Table > Auto Balloon** on ribbon.
2. Select the part list.
3. Click **OK**.
4. On the **Part List Auto-Balloon** dialog, select **Trimetric@2**.

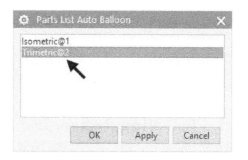

5. Click **OK** to generate balloons.

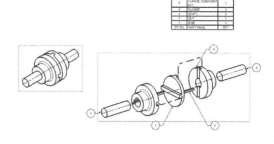

6. Save and close the file.

Chapter 5: Sketching

In this chapter, you will learn the sketching tools. You will learn to create:

- Rectangles
- Polygons
- Resolving Sketch
- Geometric Constraints
- Studio Splines
- Ellipses
- Circles
- Arcs
- Trim
- Fillets and Chamfers

TUTORIAL 1 (Creating Rectangles)

A rectangle is a four-sided 2D object. You can create a rectangle by just specifying its two diagonal corners. However, there are various methods to create a rectangle. These methods are explained next.

1. Start a new file using the **Model** template.
2. On the ribbon, click **Direct Sketch > Sketch** and select the XZ plane.
3. Click **OK** on the **Create Sketch** dialog.
4. On the ribbon, click **Direct Sketch > Rectangle** 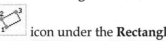.

5. Select the origin point to define the first corner.
6. Move the pointer and click to define the second corner.

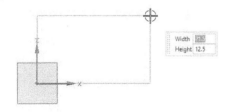

You can also type the **Width** and **Height** values to create the rectangle.

7. On the **Rectangle** toolbar, click the **By 3 Points** icon under the **Rectangle Method** section.

This option creates a slanted rectangle.

8. Select two points to define the width and inclination angle of the rectangle.

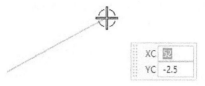

9. Select the third point to define its height.

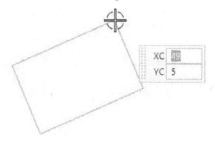

10. On the **Rectangle** toolbar, click the **From Center** icon under the **Rectangle Method** section.
11. Click to define the center point of the rectangle.
12. Move the pointer and click to define the midpoint of one side. Also, the inclination angle is defined.

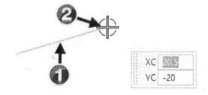

13. Move the pointer and click to define the corner point.

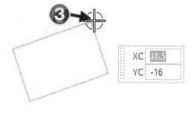

14. Click **Close** ☒ on the **Rectangle** toolbar.

Multi-Selection Gesture Drop-down

This drop-down is available on the Top Border Bar and has options to select multiple objects. The **Rectangle** option helps you to select multiple elements by dragging a rectangle covering them.

The **Lasso** option helps you to select multiple elements by dragging the pointer around them.

The **Circle** option helps you to select multiple elements by clicking and dragging a circle covering the elements.

1. On the Top Border Bar, select **Lasso** from the **Multi-Selection Gesture** Drop-down.

2. On the Top Border Bar, **Selection Scope** to **Within Active Sketch Only**.

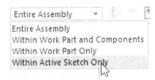

3. Press and hold the left mouse button and drag the pointer covering the slanted rectangles.

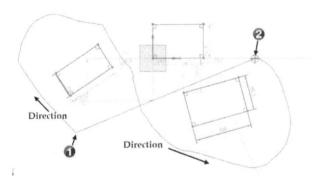

4. Press **Delete** to erase the rectangles.
5. Click **OK**.
6. Click **Undo** on the **Quick Access Toolbar**.

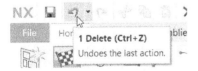

7. On the Top Border Bar, select **Circle** from the **Multi-Selection Gesture** Drop-down.
8. On the Top Border Bar, select **Curve** from the **Selection Filter** Drop-down.

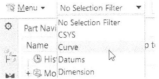

9. Click and drag a selection circle covering all the sketch elements.

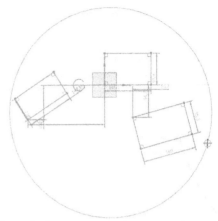

10. Press **Delete** to erase the rectangles.
11. Click **OK**.

TUTORIAL 2 (Creating Polygons)

A Polygon is a shape having many sides ranging from 3 to 513. In NX, you can create regular polygons having sides with equal length. Follow the steps given next to create a polygon.

1. Activate the **Direct Sketch** mode.
2. On the ribbon, click **Direct Sketch > Polygon**.
3. On the **Polygon** dialog, type **8** in the **Number of Sides** box under the **Sides** section.
4. Under the **Size** section, select **Size > Inscribed Radius**. This option creates a polygon with its sides touching an imaginary circle. You can also select the **Circumscribed Radius** option to create a polygon with its vertices touching an imaginary circle.
5. Click to define the center of the polygon.
6. Type **0** in the **Rotation** box.
7. Move the pointer and notice that the rotation of the polygon is constrained.
8. Type 50 in the **Radius** box and press Enter.

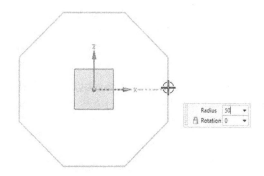

9. Click **Close** on the dialog to deactivate the tool.

Circle by 3 Points

1. On the ribbon, click **Home > Direct Sketch > Circle**.
2. On the **Circle** toolbar, click **Circle by 3 Points** icon under the **Circle Method** section.
3. Click on the vertices of the polygon. A circle is created passing through the vertices.

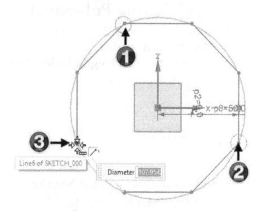

TUTORIAL 3 (Studio Splines)

Studio Splines are non-uniform curves, which are used to create irregular shapes. In NX, you can create studio splines by using two methods: **Through Points**, and **By Pole**.

1. Download the **Studio-spline example.jpg** file from:
   http://www.onlineinstructor.org/shop/nx-12-tutorial/.
2. On the ribbon, click **Home > Feature > Datum > Raster Image**.

3. On the **Raster Image** dialog, click the **Browse** icon.
4. Go to the location of the downloaded image file and double-click on it.
5. Select the XZ Plane from the Datum Coordinate System.
6. Under the **Orientation** section, select **Basepoint > Bottom Center**.
7. Set the **Reference Direction** to **Vertical**.
8. Click on the Angle handle (Spherical Dot) of the Dynamic Coordinate System, and enter 180 in the **Angle** box.

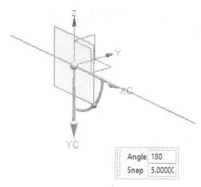

9. Expand the **Image Settings** section and set the **Overall Translucency** to 50.
10. Click **OK**.
11. Activate the **Sketch** mode on the XZ Plane.
12. On the Ribbon, click **Direct Sketch > More Curve > Studio Spline**.
13. On the **Studio Spline** dialog, select **Type > Through Points**.
14. Select the five points, as shown.

15. Select the three points, as shown.

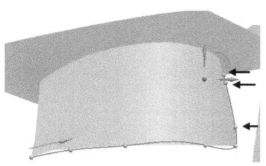

16. Likewise, select other points, as shown.

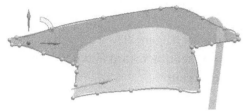

17. Under the **Parameterization** section, set the **Degree** value to 2, and check the **Closed** option.
18. Click **OK**.
19. Click **Yes** on the **Continuous Auto Dimensioning** message box. The auto dimensions are not created.

20. Double-click on the spline.
21. On the **Studio Spline** dialog, select **Type > By Poles**.
22. Drag the pole, as shown.

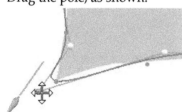

23. Likewise, modify the other pole locations, as shown.

24. Select a point on the spline, as shown. A new pole is added.

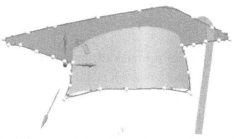

25. Drag the new pole to modify the spline.

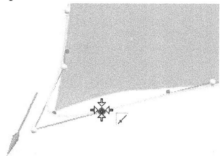

26. Likewise, add poles wherever they are required, and modify their position.
27. Click **OK**.
28. Click on the image and select **Hide**.

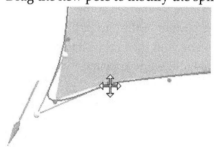

29. Click **Finish Sketch**.

TUTORIAL 4 (Geometric Constraints)

1. Activate the **Direct Sketch** mode.
2. On the ribbon, click **Home > Direct Sketch > Profile** 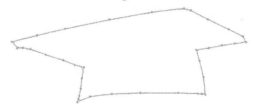.

3. Select the sketch origin.
4. Move the pointer towards right horizontally and click.
5. Move the pointer up vertically and click.
6. Move the pointer toward left and click. Notice that the Horizontal and Vertical constraints are created, automatically.

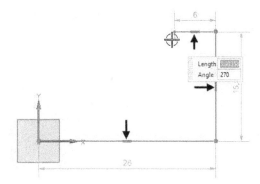

7. Move the pointer up vertically and click.
8. On the **Profile** toolbar, click the **Arc** icon.
9. Move the pointer to the end point of the previous line, and then move it toward left. An arc normal to the line appears.

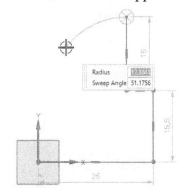

10. Move the pointer to the end point of the previous line, move toward up, and left. Notice that an arc tangent to the previous line appears.

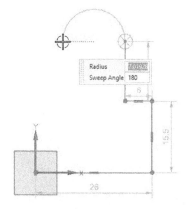

11. Click to create a tangent arc.
12. Move the pointer downward and notice the Tangent constraint.

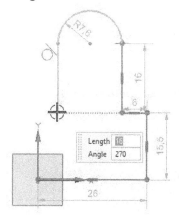

13. Click when a dotted line appears from the horizontal line.
14. Move the pointer toward left and click when a dotted line appears from the sketch origin.

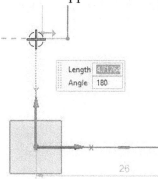

15. Move the pointer downward and click the sketch origin.

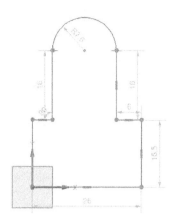

16. On the ribbon, click **Home > Direct Sketch > Circle**.

17. Select the center of the tangent arc, move the pointer outward, and click. A concentric constraint is created between the circle and the arc.

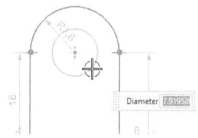

18. Place the pointer on the midpoint of the left vertical line, and move the pointer.

19. Click when a horizontal dotted line appears.

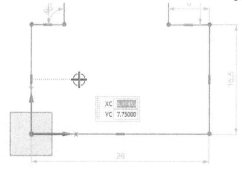

20. Move the pointer and click to create a circle.

21. Likewise, create another circle.

22. Press Esc.

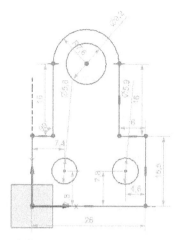

Adding Constraints

Geometric Constraints are used to control the shape of a sketch by establishing relationships between the sketch elements. You can add relations using the **Geometric Constraints** tool.

1. Select the line connected to the tangent arc, and click **Vertical** from the Shortcuts toolbar. The vertical constraint is applied to the line.

2. Select the lower vertical lines.

3. Click the **Equal Length** ＝ icon on the Shortcuts toolbar to make the lines equal.

4. Press Esc.

5. Select the two horizontal lines, as shown.

6. Click the **Equal Length** ＝ icon on the Shortcuts toolbar to make the lines equal.

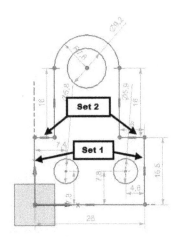

7. Press Esc.
8. Select the other two vertical lines and click the **Equal Length** = icon to make them equal.
9. Press Esc.
10. Select the two circles located at the bottom.
11. Click the **Equal Radius** = icon on the Shortcuts toolbar.
12. Select the center point of the circle and the left vertical line.
13. Click the **Midpoint** icon on the shortcuts toolbar. The midpoint of the vertical line and the center point of the circle become collinear.

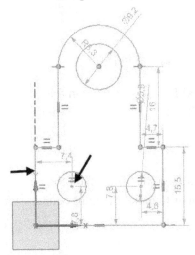

14. Likewise, make the other circle collinear with the midpoint of the right vertical line.

Adding Dimensions

1. Double-click on the dimension of the lower

right vertical line, type 30 and press Enter.

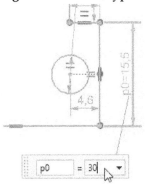

2. Likewise, change the dimension of the upper vertical line to 35.
3. On the ribbon, click **Home > Direct Sketch > Rapid Dimension**.
4. Select the lower horizontal line, move the pointer, and click to position the dimension.
5. Type 40 and press Enter.
6. Likewise, add other dimensions to the sketch to constrain it fully.

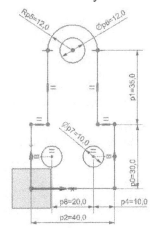

TUTORIAL 5 (Resolving Over-Constrained Sketches)

1. Activate the **Direct Sketch** mode and create the sketch, as shown.

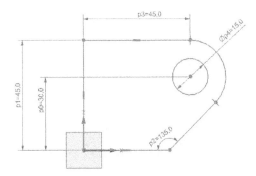

2. On the ribbon, click **Home > Direct Sketch > Rapid Dimension**.
3. Select the arc and position the dimension. The **Update Sketch** message box appears showing that the sketch is over constrained.
4. Click **OK** and press Esc. The over constrained sketch objects appear in grey color and the over-constraining dimensions and constraints appears in red.

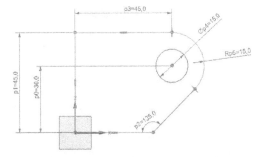

5. Select the linear dimension, as shown and click Delete.

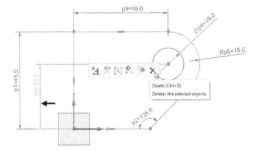

6. Click **OK** on the **Delete Dimensions** message box. Now, the sketch is fully constrained.

TUTORIAL 6 (Ellipses)

Ellipses are also non-uniform curves, but they have a regular shape. They are actually splines created in regular closed shapes.

1. Activate the **Direct Sketch** mode.
2. On the ribbon, click **Home > Direct Sketch > More Curves > Ellipse** .
3. Pick a point in the graphics window to define the location of the ellipse.
4. Type **40** and **20** in the **Major Radius** and **Minor Radius** boxes on the **Ellipse** dialog. You can also use the arrow handles to change the major and minor radius values.

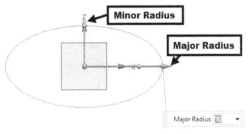

5. Type **30** in the **Angle** box. You can also use the Angle handle to rotate the ellipse. Click **OK**.

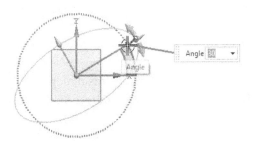

Note that the ellipse is not constrained fully. Follow the steps given next to fully-constrain the ellipse.

6. On the ribbon, click **Home > Direct Sketch > Line** .
7. Select center point of the ellipse.
8. Select a point on the ellipse.
9. Likewise, create another line.

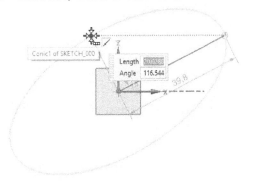

10. On the ribbon, click **Home > Direct Sketch >**

    More > Geometric Constraints .
11. Click **OK** on the message box.
12. On the **Geometric Constraints** dialog, click the

    Parallel icon under the **Constraints** section.
13. Select the line and ellipse, as shown.

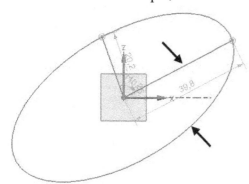

14. On the **Geometric Constraints** dialog, click the

    Perpendicular icon under the **Constraints** section.
15. Select the two lines to make them perpendicular to each other.
16. Close the **Geometric Constraints** dialog.
17. On the ribbon, click **Home > Direct Sketch >**

    Rapid Dimension .
18. Select the major axis line, move the pointer in the direction perpendicular to the line, and click to position the dimension.
19. Type 40 and press Enter.
20. Likewise, dimension the minor axis line.

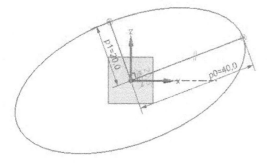

21. Select the Major axis line and the X-axis.
22. Move the pointer and click.
23. Type 30 and press Enter.

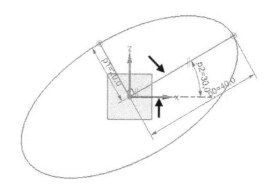

24. Close the **Rapid Dimension** dialog.
25. Click on the major axis line and select **Convert to Reference**.

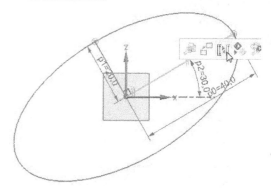

26. Likewise, convert the other line to reference.

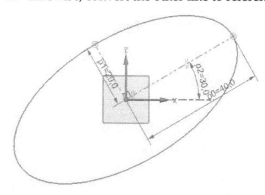

27. Double-click on the ellipse and rotate it using the Angle handle.
28. Click **OK** and notice that the ellipse returns to it position.

TUTORIAL 7 (Conics)
1. Activate the **Direct Sketch** mode.
2. Create a triangle using the **Polygon** tool.

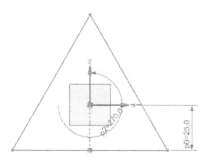

3. On the ribbon, click **Home > Direct Sketch >**

   More Curve > Conic .

4. Select the start and end limits, and control point, as shown.

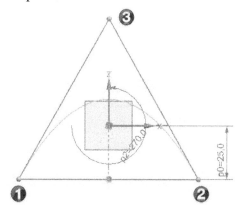

5. Set the Rho **Value** to **0.25**.
6. Expand the **Preview** section and click **Show Result**.

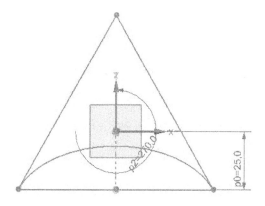

7. Click **Undo Result** on the dialog.
8. Set the Rho **Value** to **0.75** and click **Show Result**.

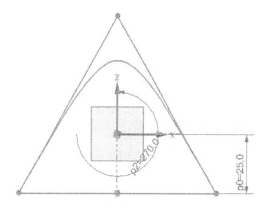

9. Click **OK**.

TUTORIAL 8 (Quick Extend, Quick Trim, Make Corner, and Offset Curve)

The **Quick Extend** tool is similar to the **Quick Trim** tool but its use is opposite of the **Quick Trim** tool. This tool is used to extend lines, arcs and other open entities to connect to other objects.

1. Create a sketch as shown below.

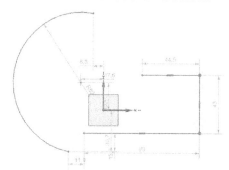

2. Click **Home > Direct Sketch > Quick Extend** on the ribbon.
3. Select the horizontal open line. This will extend the line up to arc.

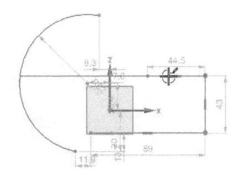

4. Likewise, extend the other elements, as shown.

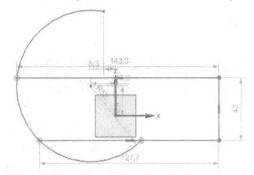

5. Close the **Quick Extend** dialog.

Make Corner

1. On the ribbon, click **Home > Direct Sketch >**

   Make Corner .
2. Click on the ending portion of the arc.
3. Click on the starting portion of the horizontal line.

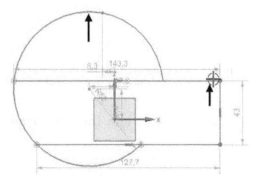

4. Close the **Make Corner** dialog.

Quick Trim

1. On the ribbon, click **Home > Direct Sketch >**

   Quick Trim .
2. Select the horizontal line.

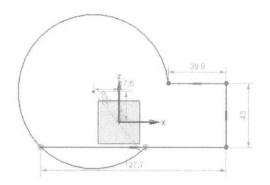

3. Close the **Quick Trim** dialog.

Offset Curve

The **Offset Curve** tool creates parallel copies of lines, circles, arcs and so on.

1. On the ribbon, click **Home > Direct Sketch >**

   More Curve > Offset Curve 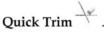 .
2. Select an entity and notice that all the connected entities are selected.
3. Type-in a value in the **Distance** box on the **Offset Curve** dialog (or) drag the arrow that appears on the offset curve.
4. Click **Reverse Direction** 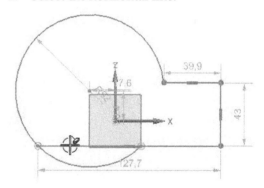 icon to reverse the offset side.
5. Click **OK**.

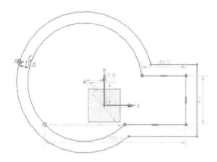

TUTORIAL 9

This tutorial teaches you to use **Fillet, Chamfer, and Mirror Curve** tools.

Fillet

The **Fillet** tool converts the sharp corners into round corners.

1. Draw the lines as shown below.

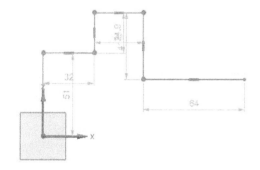

2. Click **Home > Direct Sketch > Fillet** on the ribbon.
3. Click on the corner, as shown.
4. Move the pointer and click to define the radius. You can also type the radius value.

5. On the **Fillet** toolbar, click the **Delete Third Curve** icon.

6. Select the right and left vertical lines.
7. Move the pointer and select the horizontal connecting the two vertical lines.

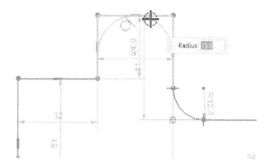

Chamfer

The **Chamfer** tool replaces the sharp corners with an angled line. This tool is similar to the **Fillet** tool, except that an angled line is placed at the corners instead of a round.

1. Click **Home > Direct Sketch > Chamfer** on the Ribbon.
2. On the **Chamfer** dialog, select **Chamfer > Symmetric.**
3. Click on the corner, as shown.

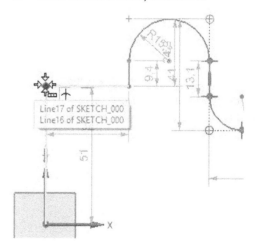

4. Type-in a value in the **Distance** box and press Enter.
5. Close the dialog.

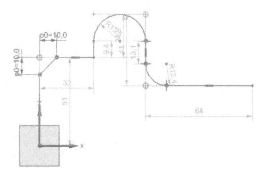

Mirror Curve

The **Mirror Curve** tool creates a mirror image of objects. You can create symmetrical sketches using this tool.

1. On the ribbon, click **Home > Direct Sketch >**

   Mirror Curve

2. Drag a selection lasso covering all the sketch entities.

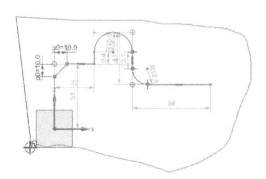

3. On the **Mirror Curve** dialog, click **Select Centerline** and select the X-axis.

4. Click **OK** to mirror the selected entities.

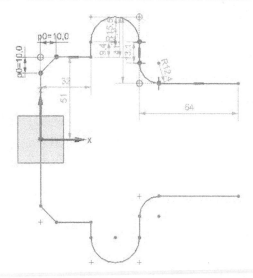

5. On the Ribbon, click **Home > Direct Sketch >**

   Arc .

6. Select the start and end points of the arc, as shown.

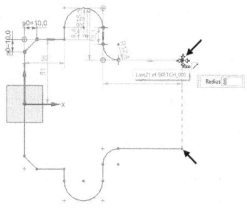

7. Move the pointer rightwards and click when the **Tangent** glyph appears.

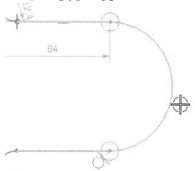

8. Close the **Arc** toolbar.

Adding Dimensions

1. Double-click on the radial dimension, as shown.

2. Type 15 and press Enter.

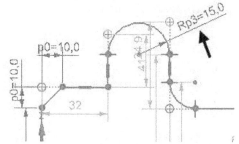

3. On the Ribbon, click **Home > Direct Sketch >**

   Rapid Dimension .

4. Select the left most vertical line and the center point of the arc, as shown.

5. Type 125 and press Enter.

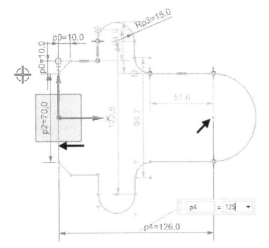

6. Select the fillet, type 5 and press Enter.
7. Likewise, create other dimensions, as shown. Deactivate the **Rapid Dimension** tool.

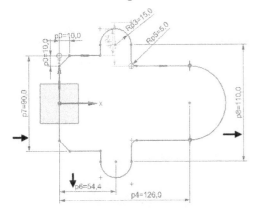

8. Zoom to the top portion of the sketch, select the points, and press Delete.

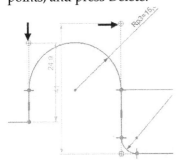

9. Click **Finish Sketch**.

TUTORIAL 10

This tutorial teaches you the **Alternate Solution** command.

1. Start a new part file.

2. On the ribbon, click **Home > Direct Sketch > Sketch**.
3. Click on the XY plane.
4. Click **OK** on the **Create Sketch** dialog.
5. Create a sketch, as shown.

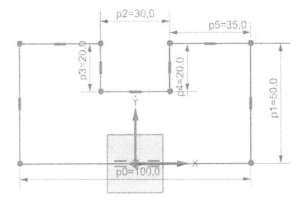

You can also download this file from our companion website.

6. On the ribbon, click **Sketch > Direct Sketch > More** gallery **> Sketch Tools > Alternate Solution** .
7. Select the vertical dimension of the sketch, as shown.

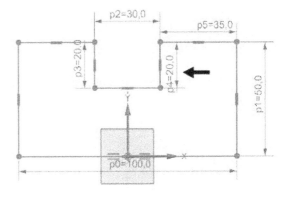

NX displays the alternate solution of the sketch.

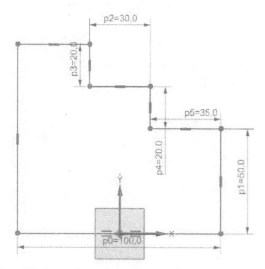

8. Click on the other vertical dimension, as shown.

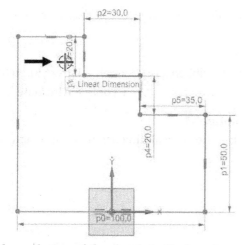

Another solution of the sketch is displayed.

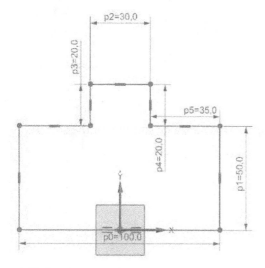

9. Click **Close** on the **Alternate Solution** dialog.
10. Click **Finish Sketch** on the ribbon.

11. Save and close the part file.

TUTORIAL 11

This tutorial teaches you the **Animate Dimension** command.

1. Create a new part file.
2. On the ribbon, click **Home tab > Direct Sketch group > Sketch**.
3. Click on the XY Plane.
4. Click **OK**.
5. Create two rectangles and add dimensions, as shown.

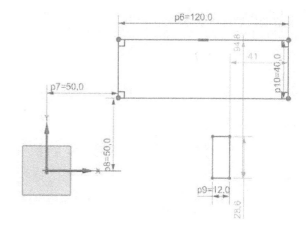

6. Select the bottom horizontal lines of the two rectangles.
7. Select **Collinear** from the Context toolbar.

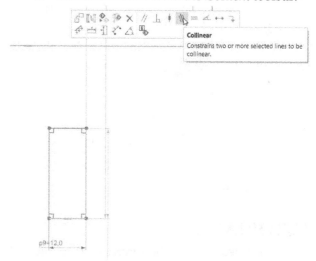

8. Likewise, make the top horizontal lines of both the rectangles **Collinear** to each other.

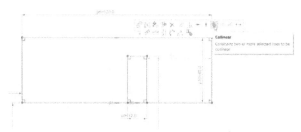

9. Click and drag the small rectangle; it freely moves between the two horizontal lines of the large rectangle.

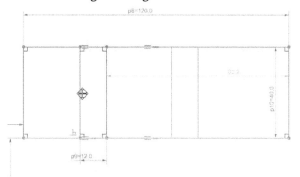

10. Activate the **Line** command.
11. Select the midpoint of the vertical line of the small rectangle, as shown.

12. Move pointer toward left, and then click to create a line.

13. Create another line connecting the end point of the previous line.

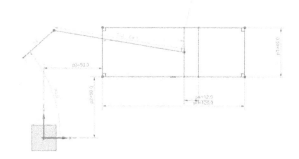

14. Create a horizontal line from the midpoint of the right vertical line of the rectangle, as shown.

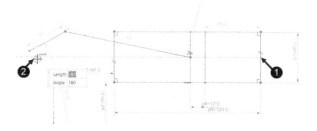

15. Click on the horizontal line, and then select **Convert to Reference** from the **Context** toolbar.

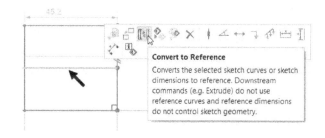

16. Select the end points of the inclined line and the reference line.
17. Select **Coincident** from the Context toolbar.

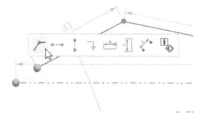

18. Add dimensions to the lines, as shown.

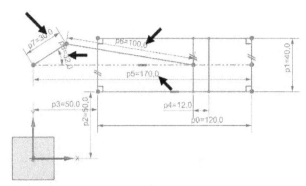

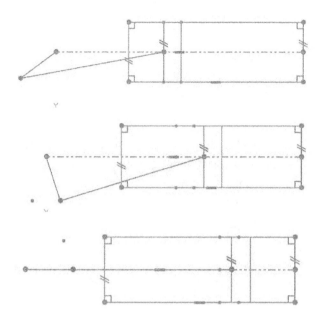

19. On the ribbon, click **Home > Direct Sketch > More gallery > Sketch Tools > Animate Dimension** .
20. Select the angular dimension from the sketch.
21. Type **0** and **360** in the **Lower Limit** and **Upper Limit** boxes, respectively.
22. Type **5** in the **Steps/Cycle** box.
23. Click the **OK** on the **Animate Dimension** dialog; the dimension value changes between 0 and 360, as shown.

24. Click **Stop** on the **Animate** message box.
25. Click **Finish Sketch** on the ribbon.

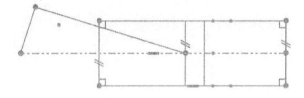

Chapter 6: Additional Modeling Tools

In this chapter, you will:

- *Construct a Sweep feature*
- *Construct a Swept feature along guide curves*
- *Create Holes*
- *Add Grooves and Slots*
- *Make Pattern Features*
- *Construct Tube features*
- *Apply Boolean operations*
- *Add chamfers*

TUTORIAL 1

In this tutorial, you will construct a helical spring using the **Helix** and **Sweep along Guide** tools.

Constructing the Helix

1. Open a NX file using the **Model** template.
2. To construct a helix, click **Curve > Curve >**

   Helix on the ribbon.
3. On the **Helix** dialog, select **Type > Along Vector**.
4. Specify the settings in the **Size** section, as given next.

5. Specify the settings in the **Pitch** section, as given next.

6. Specify the settings in the **Length** section, as given next.

7. Expand the dialog and specify the settings in the **Settings** section, as given next.

8. Click **OK** to construct the helix.

Adding the Datum Plane

1. To add a datum plane, click **Home > Feature >**
   Datum Plane on the ribbon.
2. On the **Datum Plane** dialog, select **Type > On Curve**.
3. Select the helix from the graphics window.
4. Under the **Location on Curve** section, select **Location > Through Point**.
5. Select the end of the helix.

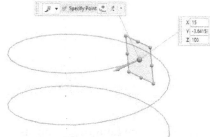

6. Under the **Orientation on Curve** section, select **Direction > Normal to Path**.
7. Leave the default values and click **OK**.

Constructing the Sweep feature

1. On the ribbon, click **Home > Direct Sketch > Sketch**.
2. Select the plane created normal to the helix.
3. Select **Origin Method > Specify Point**.
4. Click the CSYS dialog button in the **Select CSYS** section.

5. On the CSYS dialog, click in the **Specify Point** section.
6. Select the end point of the helix to define the sketch origin.

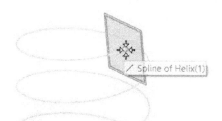

7. Click **OK** twice.
8. Draw circle of 4 mm diameter.

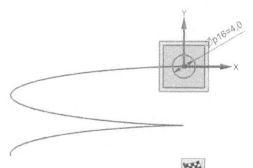

9. Right-click and select **Finish Sketch**.
10. On Top Border Bar, click the **Orient View > Isometric**.
11. To construct a sweep feature, click **Surface > More > Sweep along Guide** on the ribbon.
12. Select the circle to define the section curve.
13. Under the **Guide** section, click **Select Curve**.
14. Select the helix.
15. Leave the default settings and click **OK** to construct the sweep feature.
16. Click on the plane and select **Hide**.

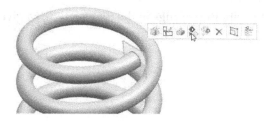

Also, hide the sketch.

17. Save and close the file.

TUTORIAL 2

In this tutorial, you construct a pulley wheel using the **Revolve** and **Groove** tools.

1. Open a file in the **Modeling** Environment.
2. Construct the sketch on the YZ plane, as shown in figure.

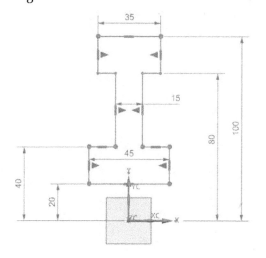

3. Finish the sketch.
4. Construct the revolved feature.

Constructing the Groove feature

1. To construct a groove feature, click **Home >**

   Feature > More > Design Feature > Groove  on the ribbon.

Note

Some tools do not appear on the ribbon. To display the required tools, select them from the menu, as shown in figure.

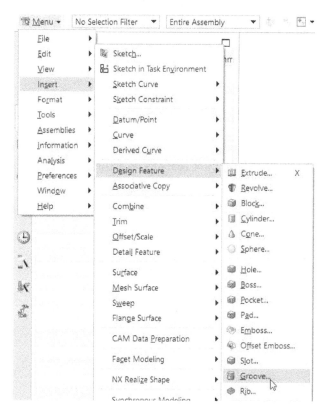

2. On the **Groove** dialog, click the **U Groove** button.

3. Select the outer cylindrical face of the revolved feature.

4. Specify the values on the **U Groove** dialog, as shown in figure.

5. Click **OK**; the **Position Groove** dialog appears.
6. Click on the cylindrical edges of the model and groove preview, as shown.

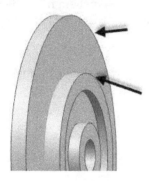

7. Enter **7.5** on the **Create Expression** dialog.

8. Click **OK** to add the groove.

9. Click **Cancel**.
10. Save and close the model.

TUTORIAL 3

In this tutorial, you construct a shampoo bottle using the **Swept**, **Extrude**, and **Thread** tools.

Creating Sections and Guide curves

To construct a swept feature, you need to create sections and guide curves.

1. Open a file in the **Modeling** Environment.
2. On the ribbon, click **Home > Direct Sketch > Sketch**.
3. Select the XY plane.
4. On the **Create Sketch** dialog, click **OK** to start the sketch.
5. On the ribbon, click **Home > Direct Sketch > Ellipse** ⊕.
6. Select the origin point of the coordinate system.
7. Specify **Major Radius** as 50 mm.
8. Specify **Minor Radius** as 20 mm.
9. Specify **Angle** as 0.
10. Leave the default settings and click **OK**.

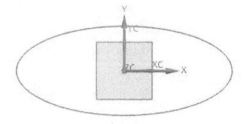

11. Fully constrain the ellipse using the procedure given Tutorial 6 of Chapter 5.

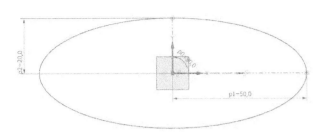

12. Click **Finish Sketch**.
13. Change the orientation to Isometric.
14. On the ribbon, click **Home > Direct Sketch > Sketch**.
15. Select the XZ plane.
16. Click **OK**.
17. On the ribbon, click **Home > Direct sketch > Studio Spline** .
18. On the **Studio Spline** dialog, select **Type > Through Points**.
19. Draw a spline similarly to the one shown in figure (refer to TUTORIAL 3 of Chapter 5: Sketching).

16. Click **OK**.
17. Select the first point of the spline and the origin point.
18. Select **Horizontal Alignment** from the Shortcuts toolbar.

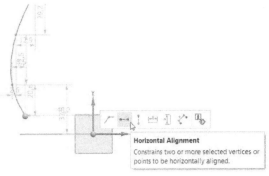

19. Apply dimensions to the spline, as shown in figure.

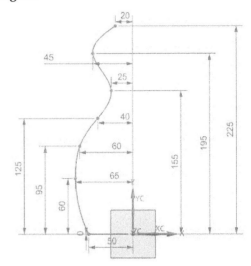

20. On the ribbon, click **Home > Direct Sketch > Mirror Curve** .
21. Select the spline.
22. On the **Mirror Curve** dialog, click **Select Centerline** and then select the vertical axis of the sketch.
23. Click **OK**.

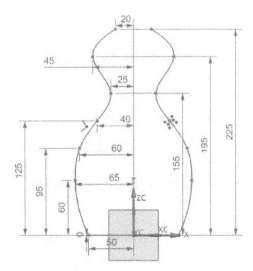

24. Click **Finish Sketch**.
25. Change the view orientation to Isometric.

Creating another section

1. On the ribbon, click **Home > Feature > Datum Plane**.
2. On the **Datum Plane** dialog, select **Type > At Distance**.
3. Select the XY plane from the coordinate system.
4. Type-in **225** in the **Distance** box.

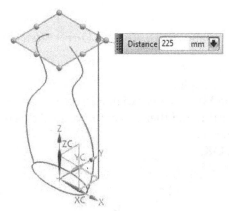

5. Click **OK**.
6. Start a sketch on the new datum plane.
7. Draw a circle of 40 mm diameter.

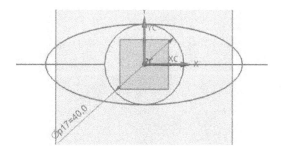

8. Click **Finish Sketch**.
9. Change the view to Isometric.

Constructing the swept feature

1. On the ribbon, click **Home > Surface > Swept**
   .
2. Select the circle and click the middle mouse button.

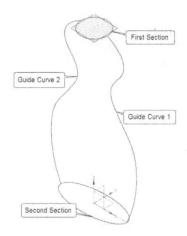

3. Select the ellipse.

Ensure that the arrows on the circle and the ellipse point towards same direction. Use the **Reverse Direction** button in the **Sections** section to reverse the direction of arrows.

8. Click **Select Curve** in the **Guides (3 maximum)** section.
9. On the Top Border Bar, select **Curve Rule > Single Curve**.

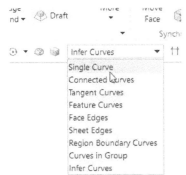

10. Select the first guide curve and click the middle mouse button.
11. Select the second guide curve.
12. Click **OK** to construct the swept feature.

Constructing the Extruded feature

1. Click on the circle on the top of the sweep feature.
2. Click **Extrude** on the contextual toolbar.

3. On the **Extrude** dialog, under the **Boolean** section, select **Boolean > Unite**.
4. Extrude the circle up to 25 mm.

Adding the Emboss feature

1. On the **Feature** group, click the **Datum Plane** button.
2. On the **Datum Plane** dialog, select **Type > At Distance**.
3. Select the XZ plane from the coordinate system.
4. Enter **50** in the **Distance** box.
5. Click **Reverse Direction** to create the plane, as shown. Click **OK**.

6. Create a sketch on the plane, as shown in figure. The major and minor radiuses of the ellipse are 50 and 20, respectively.

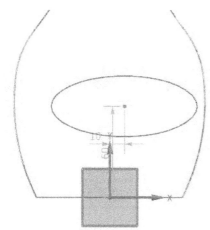

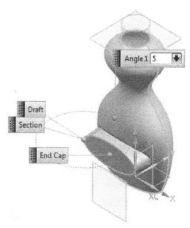

7. Select the sketch origin and the center point of the ellipse.
8. Select **Vertical Alignment** from the Shortcuts Toolbar.

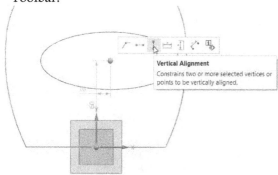

9. Apply constraints and dimensions to fully constrain the sketch (refer to TUTORIAL 6 of Chapter 5: Sketching).

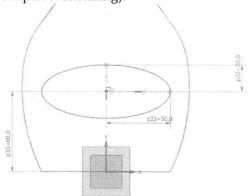

6. Click **Finish Sketch**.
7. On the ribbon, click **Home > Feature > More > Design Feature > Emboss** .
8. Select the sketch.
9. On the **Emboss** dialog, under **Face to Emboss**, click **Select Face**.
10. Select the swept feature.

11. Expand the **End Cap** section, specify the settings, as given in figure.

12. Leave the default settings and click **OK** to add the embossed feature.

Adding Edge Blend

1. On the ribbon, click **Home > Feature > Edge Blend**.
2. Click on the bottom and top edges of the swept feature.
3. Set **Radius 1** to 5 mm.

4. Click **Apply** to add the blend.
5. Set **Radius 1** to 1 mm.
6. Select the edges of the emboss feature and click **OK**.

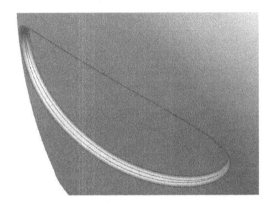

Shelling the Model

1. On the ribbon, click **Home > Feature > Shell**.
2. On the **Shell** dialog, select **Type > Remove Faces, then Shell**.
3. Set **Thickness** to 2 mm.
4. Select the top face of the cylindrical feature.

5. Click **OK** to shell the geometry.

Adding Threads

1. On the ribbon, click **Home > Feature > More > Thread** .
2. On the **Thread** dialog, set **Thread Type** to **Detailed**.
3. Select the cylindrical face.

4. Set **Pitch** to 8 mm.
5. Leave the other default settings and click **OK** to add the thread.

6. Save the model and close it.

TUTORIAL 4

In this tutorial, you construct a patterned cylindrical shell.

Constructing a cylindrical shell

1. Start a new file using the **Model** template.
2. On the ribbon, click **Home > Feature > More > Design Feature > Cylinder** .
3. On the **Cylinder** dialog, select **Type > Axis, Diameter, and Height**.
4. Select the Z-axis from the triad.

5. Specify **Diameter** and **Height** as **50** and **100**, respectively.
6. Leave the default settings and click **OK**.

7. On the ribbon, click **Home > Feature > Shell**.
8. Set **Thickness** to 3 mm.
9. Select the top and bottom faces of the cylindrical feature.
10. Click **OK** to shell the geometry.

Adding slots

1. On the ribbon, click **Home > Feature > Extrude** .
2. Click on the YZ plane; the Sketch Task Environment is displayed with the Profile command active.

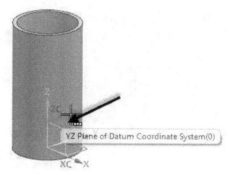

3. Click to specify the first point of the line.
4. Move the pointer upward, and then click to create a vertical line.

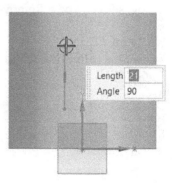

5. On the **Profile** dialog, click the **Arc** icon.
6. Move the pointer toward right, and then click.

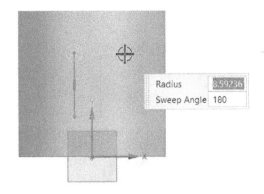

7. Move the pointer vertically downward, and then click.

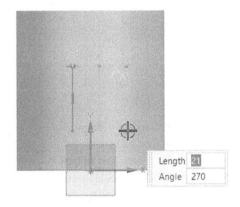

8. On the **Profile** dialog, click the **Arc** icon.
9. Select the start point of the sketch.

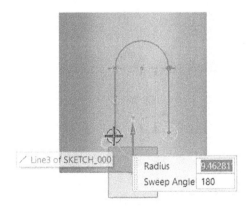

10. Press Esc to deactivate the **Profile** command.
11. Select the two arcs of the sketch, and then select the **Equal Radius** ⁓ icon on the Context toolbar.
12. Select the centerpoint of the arc and the Y axis of the sketch.
13. Select the **Point on Curve** ⌐ option from the Context toolbar.
14. Add dimensions to the sketch, as shown.

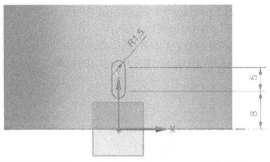

15. Click **Finish** on the **Sketch** group of the **Home** tab.
16. On the **Extrude** dialog, select **End > Until Selected** ⬚ from the **Limits** section.
17. Click on the outer cylindrical surface of the model.
18. Select the **Boolean > Subtract** from the **Boolean** section.

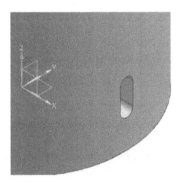

19. Click **OK** on the **Extrude** dialog.

Constructing the Linear pattern

1. On the ribbon, click **Home > Feature > Pattern Feature** ⬚ .
2. On the **Pattern Feature** dialog, select **Layout > Linear** ⬚ .
3. Select the Extrude feature.
4. Under the **Pattern Definition** section, select **Direction 1 > Specify Vector**.
5. Select the Z-axis vector.

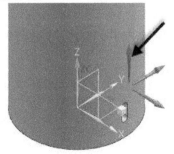

6. Select **Spacing > Count and Pitch**.
7. Type-in **6** in the **Count** box.
8. Enter **16** in the **Pitch Distance** box.
9. Click **OK** to make the linear pattern.

Constructing the Circular pattern

1. On the ribbon, click **Feature > Pattern Feature**

2. On the **Pattern Feature** dialog, select **Layout > Circular** .

3. Press and hold the Ctrl key, and then select the linear pattern and the extrude feature from the **Part Navigator**.

4. Under the **Pattern Definition** section, select **Rotation Axis > Specify Vector**.

5. Select the Z-axis vector.

Now, you have to specify the point through which the rotation axis passes.

6. Click on the circular edge of the cylindrical feature (to select the center point of the cylinder).

7. Select **Spacing > Count and Span**.
8. Type-in **12** in the **Count** box.
9. Type-in **360** in the **Span Angle** box.
10. Click **OK** to make the circular pattern.

11. Save and close the model.

TUTORIAL 5

In this tutorial, you will construct a chain.

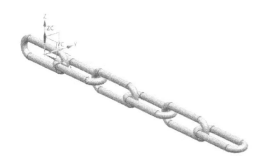

Constructing the Tube feature

1. Open a new file using the **Model** template.
2. On the ribbon, click **Home > Direct Sketch > Sketch**.
3. Select the XZ plane.
4. On the ribbon, click **Home > Direct Sketch > Profile**.
5. Click on the screen to define the first point.
6. Drag the pointer rightwards and click to define the second point.

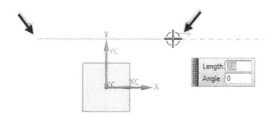

7. On the **Profile** dialog, click **Arc**.
8. Drag the mouse toward right, and then downwards.
9. Click to draw the arc.

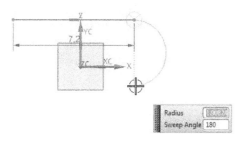

10. Drag the mouse toward left and click to define a horizontal line.

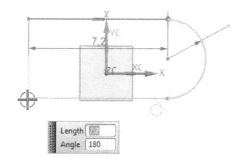

11. On the **Profile** dialog, click **Arc**.
12. Drag the mouse toward left, and then upwards.
13. Click on the start point of the sketch to draw the arc

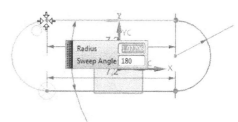

14. Close the **Profile** dialog.
15. On the ribbon, click **Home > Direct Sketch > More > Make Symmetric**.
16. Select the two arcs and click on the vertical axis.

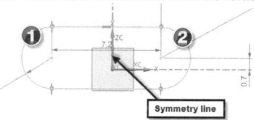

17. Click **Reset** ⟳ on the **Make Symmetric** dialog.
18. Select the horizontal lines, and then click on the horizontal axis.
19. Add dimensions to the sketch.

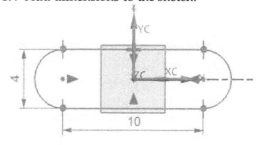

20. Click **Finish Sketch** on the **Direct Sketch** group.
21. To construct a tube feature, click **Home > Surface > More > Tube** ⬭ on the ribbon.
22. Select the sketch.
23. On the **Tube** dialog, type-in 1.5 and 0 in the

Outer Diameter and **Inner Diameter** boxes, respectively.

24. Click **OK** to construct the tube feature.

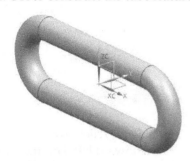

Patterning the Tube geometry

1. On the ribbon, click **Home > Feature > Pattern Feature**.
2. On the **Pattern Feature** dialog, select **Layout > Linear** and click on the tube feature.
3. Under the **Pattern Definition** section, select **Direction 1 > Specify Vector**.
4. Select the X-axis vector.

5. Under **Direction 1**, select **Spacing > Count and Pitch**.
6. Type-in **6** and **12** in the **Count** and **Pitch** boxes, respectively.
7. Expand the **Orientation** section and select **Orientation > CSYS to CSYS**.
8. Under **Orientation**, select **Specify From Vector CSYS > CSYS Dialog** .
9. On the **CSYS** dialog, select **Type > Dynamic**.
10. Accept the default position of the Dynamic CSYS and click **OK**.

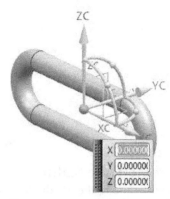

11. On the **Pattern Feature** dialog, under **Orientation**, select **Specify To CSYS > CSYS Dialog**.
12. Rotate the Dynamic CSYS about the X-axis. The rotation angle is -90 degrees.

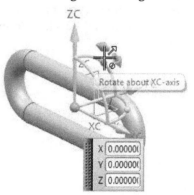

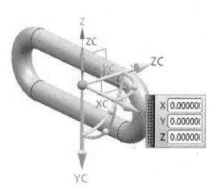

13. Click **OK**.
14. On the **Pattern Feature** dialog, under **Orientation**, check the **Repeat Transformation** option.
15. Click **OK** to make pattern of the tube.

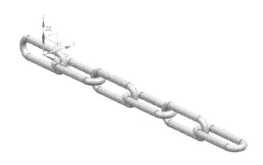

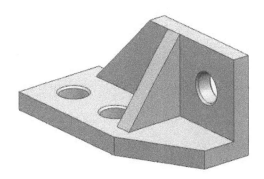

16. Save and close the file.

Boolean Operations

Types of Boolean operations.

Unite
Subtract
Intersect

These tools combine, subtract, or intersect two bodies. Activate these tools from the **Combine** drop-down on the **Feature** group.

Unite: This tool combines the **Tool Body** and the **Target Body** into a single body.

Subtract: This tool subtracts the **Tool body** from the **Target body**.

Intersect: This tool keeps the intersecting portion of the tool and target bodies.

TUTORIAL 6

In this tutorial, you will construct the model shown in figure.

Constructing the first feature

1. Open a new part file.
2. Construct the first feature on the XY plane (create a rectangular sketch and extrude it up to a distance of 10 mm).

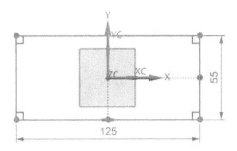

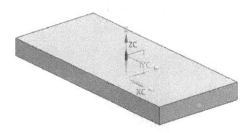

Constructing the Second Feature

1. Draw the sketch on the top face of the first feature.

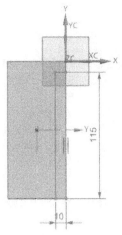

2. On the ribbon, click **Home > Feature > Extrude**.
3. Select the sketch.
4. Type-in **45** in the **End** box.

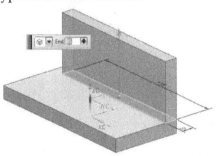

5. Under the **Boolean** section, select **Boolean > Unite**.
6. Click **OK**.

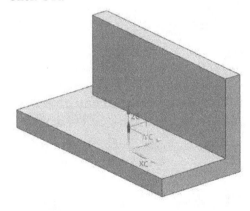

Constructing the third feature

1. On the ribbon, click **Home > Feature > Datum Plane**.
2. On the **Datum Plane** dialog, select **Type > At Distance**.
3. Click on the right-side face of the model geometry.
4. Type-in **50** in the **Distance** box and click the

Reverse Direction ⊠ icon on the **Datum Plane** dialog.

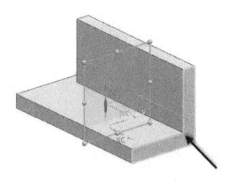

5. Click **OK**.
6. Draw the sketch on the new datum plane.

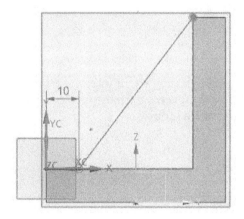

7. On the ribbon, click **Home > Feature > More > Design Feature > Rib** .
8. Select the sketch.
9. On the **Rib** dialog, select **Walls > Parallel to Section Plane**.
10. Under the **Walls** section, select **Dimension > Symmetric** and type-in **10** in the **Thickness** box.
11. Check **Combine Rib with Target**.
12. Click **OK**.

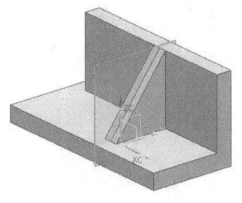

Drilling Holes

1. To drill holes, click **Home > Feature > Hole** 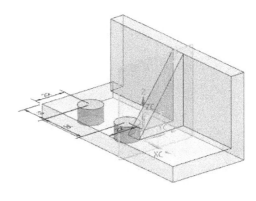 on the ribbon.
2. On the **Hole** dialog, select **Type > Drill Size Hole**.
3. Under the **Forms and Dimensions** section, select **Size > 16**.
4. Select **Depth Limit > Through Body**.
5. Expand the **Start Chamfer** and **End Chamfer** sections, and then check the **Enable** option.
6. Click on the top face of the model.

11. Click **Apply** to create the hole.

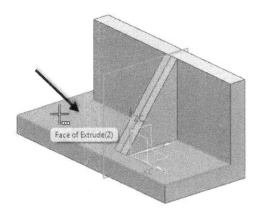

7. Click to place one more point.
8. Click **Close** on the **Sketch Point** dialog.
9. Add dimensions to define the hole location.

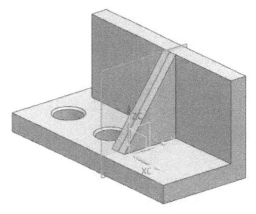

12. Drill another hole on the front face of the second feature.

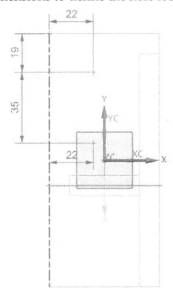

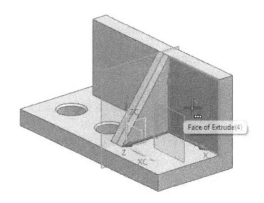

10. Click **Finish** on the ribbon.

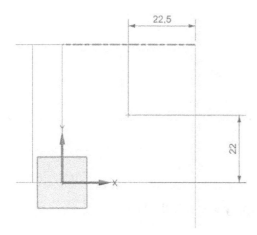

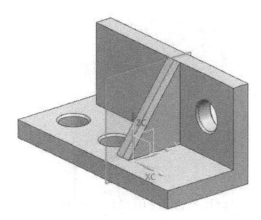

Adding Chamfers

1. To add a chamfer, click **Home > Feature > Chamfer** on the ribbon.
2. On the **Chamfer** dialog, select **Cross-section > Asymmetric**.
3. Under the **Offsets** section, type-in **25** and **45** in the **Distance 1** and **Distance 2** boxes.
4. Click on the corner edge of the first feature.

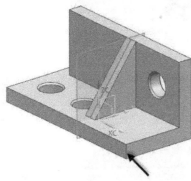

5. Click **Apply** add the chamfer.

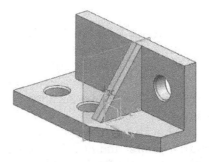

6. On the **Chamfer** dialog, select **Cross Section > Symmetric**.
7. Type-in **45** in the **Distance** box.
8. Click on the corner edge of the second feature.

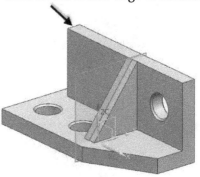

9. Click **OK**.

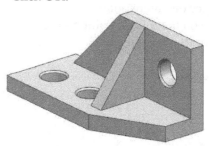

10. Save the model.

Edit Parameters

1. Click on the Drilled hole and select **Edit Parameters** from the Shortcuts toolbar.

2. On the **Hole** dialog, select **Type > General Hole**.
3. Under the **Form and Dimensions** section, select **Form > Counterbored**.

4. Set the Dimensions of the counterbored hole, as shown.

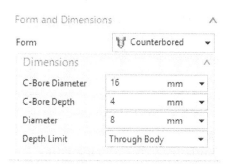

You can also change the location of the hole by double clicking on the location dimensions and changing their values.

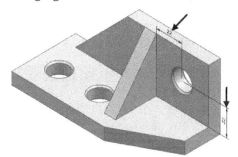

5. Click **OK**.

Show Dimensions

1. Click on the base feature and select **Show Dimensions** from the Shortcuts toolbar.

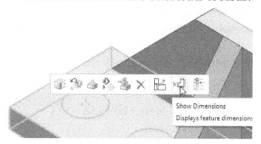

2. Double click on the linear dimension of the extrude feature.

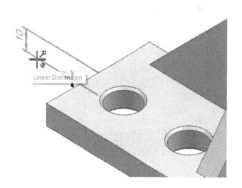

3. On the **Feature Dimension** dialog, type-in 20 in the value box and click **OK**.

4. Right click and select **Refresh** or press F5.

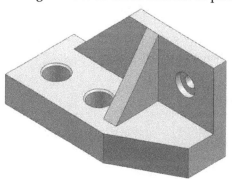

Editing Features by Double-clicking

1. Double-click on the chamfer.

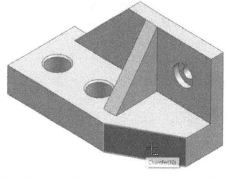

5. On the **Chamfer** dialog, type-in 30 in the **Distance 1** box and click **OK**.

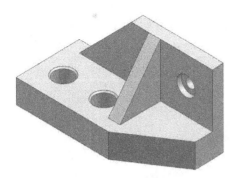

Supress Features

1. Click on the chamfer face and select **Suppress** on the Shortcuts toolbar.

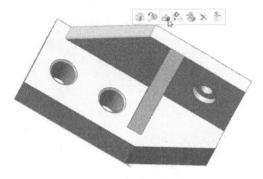

2. On the Part Navigator, check the **Chamfer** feature to unsuppress it.

3. On the Part Navigator, right click on **Extrude (1)** (base feature), and then select **Make Sketch External**.

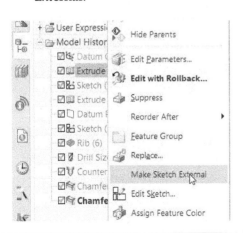

4. On the Part Navigator, right click on the Sketch of the base feature and select **Edit Parameters**.
5. On the **Edit Sketch Dimensions** dialog, select the 125 dimension and change its name to **Length**.

6. Click **Apply** and **OK**.
7. On the Top Border Bar, click **Menu > Edit > Feature > Suppress by Expression**.
8. Select the previously unsuppressed chamfer and click **Apply**.
9. Click **Show Expressions** on the dialog. The **Information** window appears showing the chamfer expression. The value 1 indicates that it is currently unsuppressed.

10. Close the **Information** window and the **Suppression By Expression** dialog.
11. On the ribbon, click **Tools > Utilities > Expression**.
12. On the **Expressions** dialog, select **Show > All Expressions**.
13. Scroll down and select **p278 (Chamfer (11) Suppression Status)** from the listed expressions.

Note: The expression number may differ in your case.

14. Double click in the **Name** box and enter **Chamfer_Suppression**.
15. Enter **if (Length=>125) (1) else (0)** in the **Formula** box.
16. Click **OK**.
17. On the Part Navigator, right click on the Sketch of the base feature and select **Edit Parameters**.

18. On the **Edit Sketch Dimensions** dialog, select the 125 dimension and change its value to **124**.

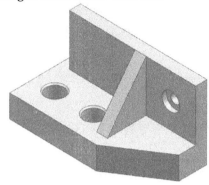

19. Click **OK** . The chamfer is suppressed as the length value is less than 125.

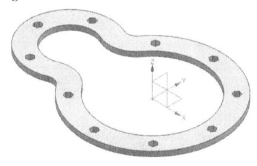

20. Close the file.

TUTORIAL 7

In this tutorial, you create the model shown in figure.

5. On the ribbon, click **Home > Curve > Quick Trim** and trim the intersecting entities.

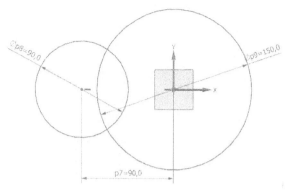

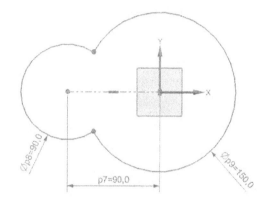

6. On the ribbon, click **Home > Curve > Fillet** and set the **Radius** value to 10.
7. Select the intersecting corners of the circles.

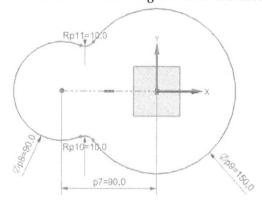

Constructing the first feature

1. Open a new part file.
2. On the ribbon, click **Home > Features > Extrude**.
3. Click on the XY plane.
4. Construct two circles and add dimensions to them.

8. Click **Finish**.
9. Extrude the sketch up to 5 mm distance.

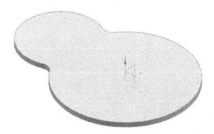

Constructing the Extruded cut

1. On the ribbon, click **Home > Feature > Extrude**.
2. Click on the top face of the model.
3. On the ribbon, click **Home > Curve > Offset Curve** .
4. Click on any edge of the top face.
5. On the **Offset Curve** dialog, set the **Distance** value to **20**.
6. Click the **Reverse Direction** button.
7. Click **OK**.

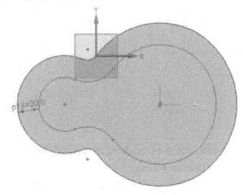

8. Click **Finish**.
9. On the **Extrude** dialog, click the **Reverse Direction** button under the **Direction** section.
10. Click **OK**.

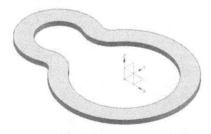

Constructing the Extruded cut

1. On the ribbon, click **Home > Feature > Extrude**.
2. Click on the top face of the model geometry.
3. On the ribbon, click **Home > Curve > Polygon** .

4. On the **Polygon** dialog, type-in **6** in the **Number of Sides** box.
5. Select **Size > Circumscribed Radius**.
6. Set the **Radius** to **4**.
7. Set the **Rotation** to **0**.
8. Click to define the center point of the polygon.

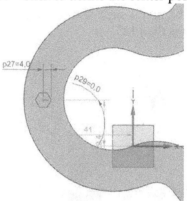

9. Close the **Polygon** dialog.
10. Select the center point of the polygon and the origin.
11. Select **Horizontal Alignment** from the **Shortcuts Toolbar**.

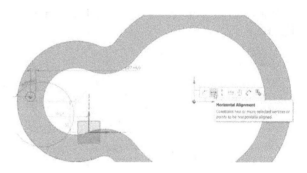

12. Add the dimension between the center point of the polygon and the origin.

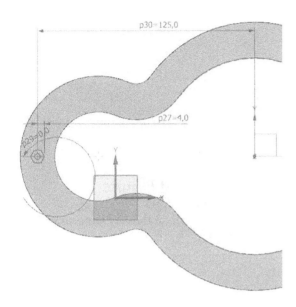

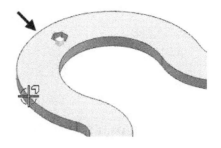

7. Select the outer edge of the top face.
8. Select **Spacing > Count and Span**.
9. Type-in **10** in the **Count** box.
10. Type-in **100** in the **% Span By** box.
11. Under the **Orientation** section, set **Orientation** to **Normal to Path**.
12. Click **OK**.

13. Click **Finish**.
14. Create the cut throughout the body.

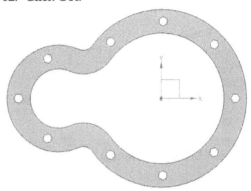

Making the Along Pattern

1. On the ribbon, click Home > **Feature > Pattern Feature** .
2. Select the polygonal cut to define the feature to pattern.
3. Select **Layout > Along**.
4. Select **Path Method > Offset**.
5. Click **Select Path**.
6. On the Top Border Bar, select **Curve rule > Tangent Curves**.

Measuring the Mass Properties

1. On the ribbon, click **Analysis > Measure > More > Measure Body** .
2. Select the geometry. Notice the volume of the geometry. You can select a different property from the drop-down to see its value.

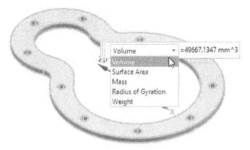

You can also check the **Show Information Window** option under the **Results Display** section to display the mass properties in the **Information** window.

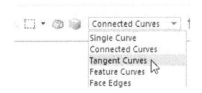

3. Click **Cancel** on the **Measure Body** dialog.
4. Close the **Information** window, if opened.
5. On the Top Border Bar, click **Menu > Edit > Feature > Solid Density** .
6. On the **Assign Solid Density** dialog, select **Units > lbm/in3**.
7. Type **0.45** in the **Solid Density** box.
8. Select the geometry and click **OK**.
9. On the ribbon, click **Analysis > Measure > More > Measure Body** .
10. Select the geometry and notice the updated mass properties.
11. Click **Cancel**.
12. On the ribbon, click **Tools > Utilities > More > Assign Materials** .
13. Select the geometry.
14. Select **Iron_Malleable** from the **Materials** section.
15. Click **Apply**.
16. Right click on the **Iron_Malleable** material and select **Inspect** from shortcut menu. The **Isotropic Material** dialog appears showing various properties of the material. You can view the Mechanical, Strength, Durability, Formability and other properties by clicking on each of them.
17. Close the **Isotropic Material** dialog and click **OK**.
18. Use the **Measure Body** tool to see the Mass Properties of the geometry.
19. Save and close the file.

TUTORIAL 8

In this tutorial, you create a plastic casing.

Creating the First Feature

1. Open a new part file.
2. Create a sketch on the XY Plane, as shown in figure.

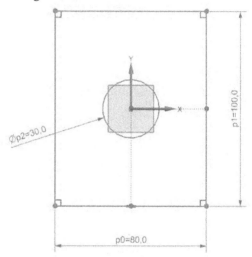

3. Click **Finish Sketch** .
4. Click the **Feature > Extrude** on the ribbon.
5. Select the sketch.
6. Set the **Distance** to 30.
7. Expand the **Draft** section and select **Draft > From Start Limit**.
8. Set the **Angle** to **2 deg**.
9. Click **OK**.

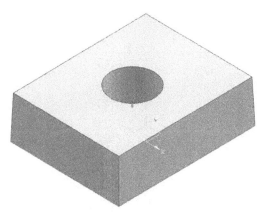

Creating the Extruded surface

1. Click **Feature > Extrude** on the ribbon and select the YZ Plane.
2. Create a sketch, as shown. Note that there should a Tangent constraint between the arc and the inclined edge of the geometry.

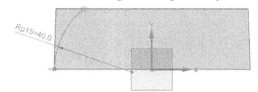

3. Click **Finish**.
4. On the dialog, under the **Limits** section, select **Start > Symmetric Value**.
5. Type-in **50** in the **Distance** box and click **OK**.

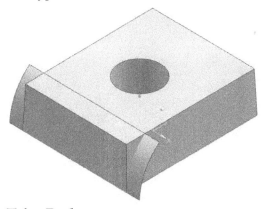

Trim Body

1. On the ribbon, click **Feature > Trim Body**.

Now, you need to select the target body.

2. Select the solid body.

Next, you need to select the tool body.

3. Select **Tool Option > Face or Plane**.
4. Click **Select Face or Plane** and select the extruded surface.
5. Make sure that the arrow points towards front. You can double-click on it to reverse its direction.
6. Click **OK** to trim the solid.

7. Hide the extruded surface by clicking on it and selecting **Hide**.

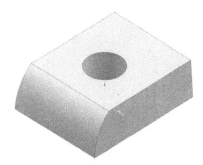

Variable Radius Blend

1. On the ribbon, click **Home > Feature > Edge Blend**.
2. Select the edge between the top and curved faces.

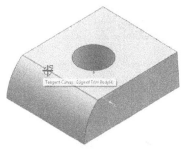

3. Expand the **Variable Radius** section and click **Specify Radius Point**.
4. Select the three points on the edge, as shown.

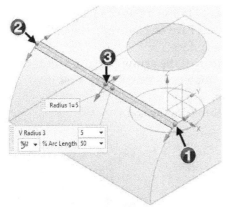

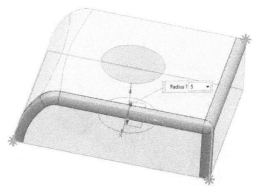

5. Under the **Variable Radius** section, expand the **List** section.
6. Select the radius points one-by-one and change the radius values, as shown.

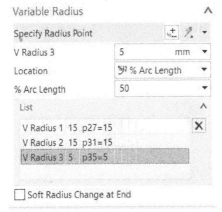

4. Expand the **Corner Setback** section and click **Select End Point**.
5. Select the vertex point, as shown.

7. Click **OK** to create the variable radius blend.

6. Under the **Corner Setback** section, expand the **List** section.
7. Select the setback points one-by-one and change the setback values, as shown. You can also change the setback values on the handle attached to the corner.

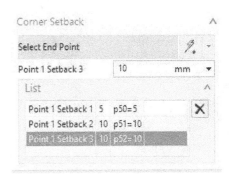

Corner Setbacks

1. On the ribbon, click **Home > Feature > Edge Blend** .
2. Set the Radius 1 to 5
3. Select the edges of the geometry, as shown.

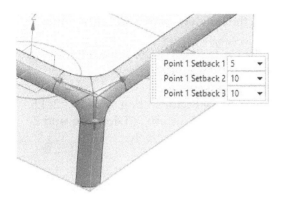

8. Under the **Edge** section, click **Select Edge**.
9. Select the edges, as shown.

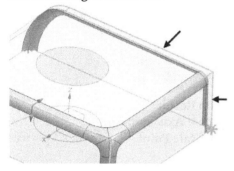

10. Under the **Corner Setback** section, click **Select End Point**.
11. Select the vertex point, as shown.

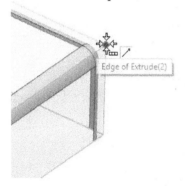

12. Change the setback values on the handle attached to the corner.

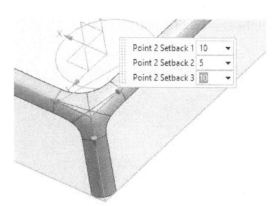

13. Click **OK**.

Shell with an Alternate Thickness

1. On the ribbon, click **Home > Feature > Shell**.
2. Select the bottom face of the geometry.
3. Under the **Thickness** section, type-in 2 in the **Thickness** box.
4. Expand the **Alternate Thickness** section and click **Select Face**.
5. Select the cylindrical face, as shown.
6. Under the **Alternate Thickness** section, type-in 4 in the **Thickness** box.

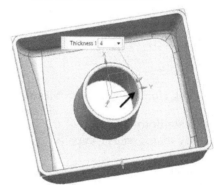

7. Click **OK**.

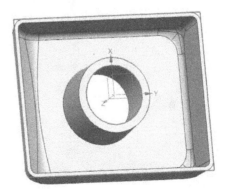

Offset Face

1. On the ribbon, click **Home > Feature > More > Offset/Scale > Offset Face**.

2. Select the face of the geometry, as shown.

3. Drag the arrow handle downward, and release when the offset value is set to 20.

4. Click **OK**.

Scale Body
1. On the ribbon, click **Home > Feature > More > Offset/Scale > Scale Body**.
2. On the **Scale Body** dialog, select **Type > Uniform**.
3. Select the geometry.
4. Under the **Scale Point** section, click **Specify Point**.
5. Select the center point of the circular edge, as shown.

6. Type-in **1.05** in the **Uniform** box.

8. Expand the **Preview** section and click **Show Result**.
7. Click the **Undo Result** button in the **Preview** section.
8. Select **Type > General**.
9. Under the **Scale Factor** section, type-in 1.2, 1.5, and 0.8 in the **X Direction**, **Y Direction**, and **Z Direction** boxes, respectively.
10. Click **Show Result**.

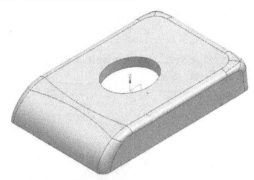

11. Click **Undo Result**.
12. Select **Type > Axisymmetric**.
13. Under the **Scale Point** section, click **Specify Vector**.
14. Select the Z axis from the triad.
15. Click **Specify Axis Through-point** and select the center point of the circular edge, as shown.
16. Under the **Scale Factor** section, type-in 2 in the **Along Axis** box.
17. Click **Show Result**.

18. Click **Cancel**.

Extract Geometry
1. On the ribbon, click **Home > Feature > More > Associate Copy > Extract Geometry**.
2. On the **Extract Geometry** dialog, select **Type > Composite Curve**.
3. Select the edges of the geometry, as shown.

4. Click **OK**.
5. On the Part Navigator, right click on the **Composite Curve** and select **Hide Parents**. The geometry is hidden.

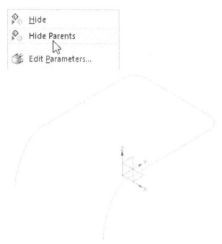

6. Right click on the Composite Curve and select **Show Parents**.
7. On the ribbon, click **Home > Feature > More >** **Associate Copy > Extract Geometry** .
8. On the **Extract Geometry** dialog, select **Type > Face**.
9. Select the top face of the geometry and click **OK**.
10. In the Part Navigator, right click on the extracted face and select **Hide Parents**.

11. Right click on the extracted face and select **Show Parents**.

12. On the ribbon, click **Home > Feature > More >** **Associate Copy > Extract Geometry** .
13. Select **Type > Region of Faces**.
14. Select the inner horizontal face of the shell feature.

15. Select the thin planar face to define the boundary.

16. Under the **Region Options** section, check the **Traverse Interior Edges** option.
17. Expand the **Preview** section and click **Preview Region**.

18. Click **Finished Preview**.
19. Uncheck the **Traverse Interior Edges** option.
20. Click **Preview Region**.

21. Click **OK**.
22. Close the file.

TUTORIAL 9

In this tutorial, you will learn the **Reorder Feature**, and **Replace Feature** tools.

1. Download and open the Tutorial 9 file.

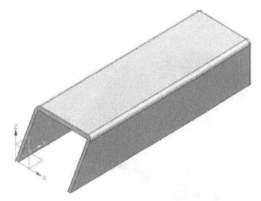

Notice that the edge blend is applied only on the outside edges of the geometry.

2. In the Part Navigator, click the Edge Blend, drag it, and place above the Shell.

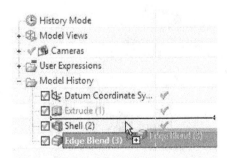

The edge blend is applied to both the inner and outer edges of the shell feature, automatically.

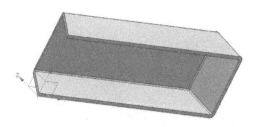

Replace Features

1. In the Part Navigator, right click on the **Extrude** feature and select **Make Current Feature**.
2. Download the Tutorial 9-Replacement.
3. Click **File > Import > Part**.
4. On the **Import Part** dialog, uncheck the **Create Named Group** option, leave the default settings, and click **OK**.
5. Browse to the location of the Tutorial 9-Replacement part and double-click on it.
6. On the **Point** dialog, leave the X, Y, and Z values to 0 and click **OK**.

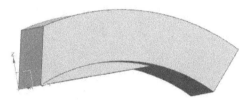

7. Click **Cancel**.
8. In the Part Navigator, right click on the **Shell** feature and select **Make Current Feature**.
9. In the Part Navigator, right click on the first **Extrude** feature and select **Replace**.
10. On the **Replace Feature** dialog, click **Select Feature** under the **Replacement Feature** section.
11. Select the imported geometry.
12. Click the **Next** button in the **Mapping** section until the back edge of the extrude feature is highlighted.

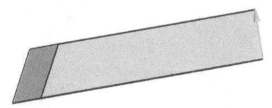

13. Select the corresponding edge on the replacement feature.

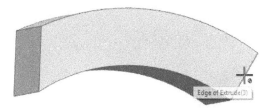

14. Likewise, select the corresponding references on the replacement feature.
15. Expand the **Settings** section and check the Delete Original Feature option.
16. Click **OK**.

17. Save and close the file.

TUTORIAL 10

In this tutorial, you will learn to divide faces, and apply draft using the **To Parting Edges** option.

1. Download and open the Tutorial 10 file.

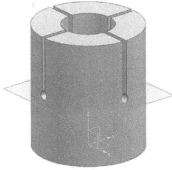

2. On the ribbon, click **Home > Feature > More > Trim > Divide Face** .
3. Select the outer cylindrical face of the geometry.
4. Click **Select Object** under the **Dividing Objects** section.
5. Select the datum plane.
6. Leave the **Projection Direction** to **Normal to Face**.
7. Click **OK**. The cylindrical face is divided into two parts.

Applying Draft using the Parting Edge option

1. On the ribbon, click **Home > Feature > Draft** .
2. On the **Draft** dialog, select **Type > Parting Edge**.
3. Click **Select Plane** under the **Stationary Plane** section.
4. Select any point on the parting edge, as shown.

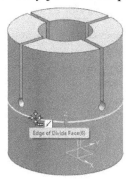

5. Click **Select Edge** under the **Parting Edges** section.
6. Select the parting edge and enter 10 in the **Angle 1** box.

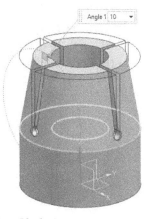

7. Click **OK**.
8. Save and close the file.

TUTORIAL 11

In this tutorial, you will learn to apply draft using the **Face** option.

1. Download and open the Tutorial 11 file.

2. On the ribbon, click **Home > Feature > Draft**.

3. On the **Draft** dialog, select **Type > Face**.
4. Select the Z-axis from triad to define the drafting direction.
5. Under the **Draft References** section, select **Draft Method > Parting Face**.
6. Click **Select Stationary Parting Face** under the **Draft References** section.
7. Select the parting surface, as shown.

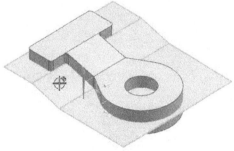

8. Click **Select Face** under the **Faces to Draft** section.
9. On the Top Border Bar, set the **Face Rule** to **Tangent Faces**.
10. Select anyone of the tangentially connected faces, as shown.

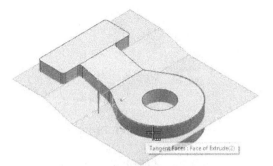

11. Type **10** in the **Angle 1** box.
12. Check the **Draft Both Sides** option under the **Draft References** section.
13. Uncheck the **Symmetric Angle** option under the **Faces to Draft** section.
14. Type **15** in the **Below Angle 1** box.

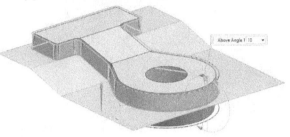

15. Click **OK**.
16. Click on the parting surface and select **Hide**.

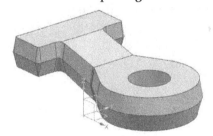

17. Save and close the file.

TUTORIAL 12

In this tutorial, you will learn to apply draft using the **Tangent to Face** option.

1. Download and open the Tutorial 12 file.

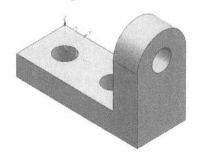

2. On the ribbon, click **Home > Feature > Draft** .

3. On the **Draft** dialog, select **Type > Tangent to Face**.

4. Select the Z-axis from triad to define the drafting direction.

5. Select the cylindrical face. The faces connected tangentially to the cylindrical face are drafted

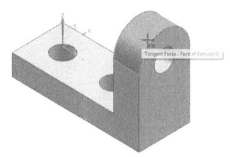

6. Type **10** in the **Angle 1** box and click OK.

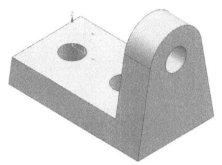

7. Save and close the file.

TUTORIAL 13

In this tutorial, you will learn to create Feature groups.

1. Download and open the Tutorial 13.

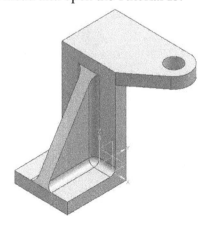

2. On the Top Border Bar, click **Menu > Format > Group > Feature Group** .

3. Type **Rib_with_blends** in the **Feature Group Name**.

4. Press the Ctrl key and select **Rib (3)** and **Edge Blend (4)** from the **Features in Part** list.

5. Click the **Add** icon to add them to the **Features in Group** list.

6. Click **OK**. The feature group appears in the **Part Navigator**.

7. Uncheck the **Feature Group** option in the Part Navigator. The **Feature** group is suppressed.

8. Check the **Feature Group** option to unsuppress it.

9. Save and close the file.

TUTORIAL 14

In this tutorial, you will learn to create a Swept Volume feature.

1. Open new NX file using the Model template.

2. Create a cylindrical feature of 40 mm diameter and 50 mm height.

3. On the ribbon, click **Curve > Curve > Helix** .

4. Click the Angle handle on the CSYS manipulator, and then type 90 in the Angle box.

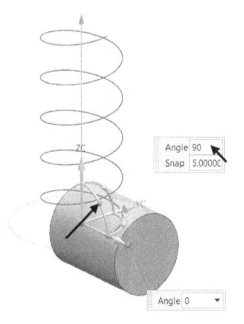

5. On the **Helix** dialog, under the **Size** section, click the down arrow next to the **Value** box, and then select **Measure**.

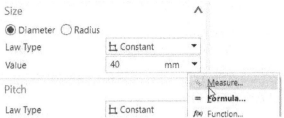

6. On the **Measure Distance** dialog, select **Type >
   Diameter**.
7. Select the cylindrical surface, and then click **OK**.

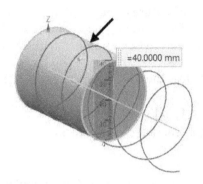

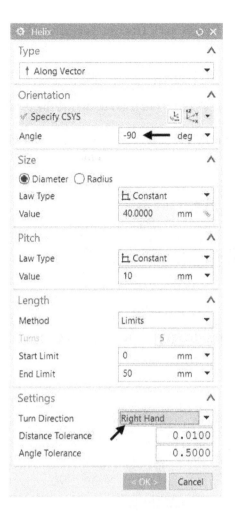

8. Enter **10** in the **Value** box of the **Pitch** section.
9. Select **Limits** from the **Method** drop-down.
10. Enter **50** in the **End Limit** box of the **Length**
    section.
11. Enter **-90** in the **Angle** box of the **Orientation**
    section.
12. Under the **Settings** section, select **Turn
    Direction > Right Hand**.

13. Click **OK**.

Creating the Tool Body

1. On the ribbon, click **Home > Direct Sketch >
   Sketch**.
2. Select the XY Plane from the Datum coordinate
   system, and then click **OK**.
3. Create a sketch, as shown in figure below.

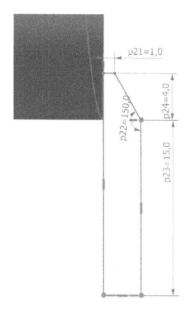

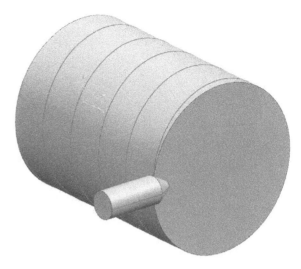

4. Click **Finish Sketch** on the ribbon.
5. On the ribbon, click **Home > Feature > Extrude > Revolve.**
6. Select the sketch from the graphics window.
7. On the **Revolve** dialog, click **Specify Vector** in the **Axis** section.
8. Select the vertical line of the sketch, as shown.

Creating the Swept Volume Feature

1. On the ribbon, click **Home > Feature > More > Sweep > Swept Volume** .
2. Select the Tool body from the graphics window; the target body is selected automatically.
3. Select the helix.
4. Under the **Orientation** section, select **Sweep Orientation > Normal to Path.**
5. Select the X-axis from the triad.

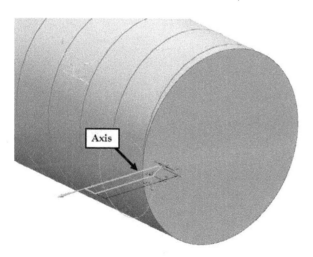

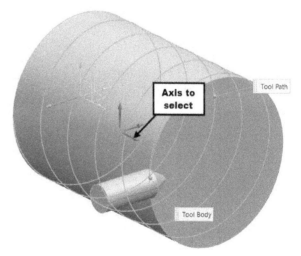

9. Enter 360 in the **End** box attached to the sketch.
10. Under the **Boolean** section, select **Boolean > None**; a separate solid is created.
11. Click **OK**.

6. Make sure that the **Boolean** type is set to **Subtract**.
7. Click **OK** to create the swept volume feature.

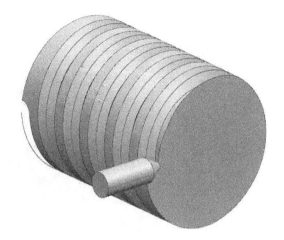

Chapter 7: Expressions

In this chapter, you will:

- *Use Program Generated Expressions*
- *Create your own Expressions*
- *Create Family of Parts*
- *Create expressions by measuring elements*
- *Export and Import Expressions*

TUTORIAL 1

In this tutorial, you will modify the basic expressions which are automatically created by NX.

1. Start a new file in the Modeling environment.
2. Activate the **Direct Sketch** mode on the XY plane.
3. Create the sketch, as shown.

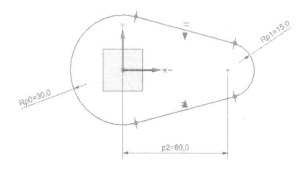

Notice the expressions that are applied to the dimensions.

4. Click **Finish Sketch** on the ribbon.
5. Click **Orient View Drop-down > Isometric** on the Top Border Bar (or) press the End key.
6. On the ribbon, click **Home > Feature > Extrude**.
7. Select the sketch.
8. On the **Extrude** dialog, set the values in the **Limits** section, as given below:

   Start: Value
   Distance: 0
   End: Value
   Distance: 15

9. Expand the **Offset** section and set the values, as given below:

Offset: Symmetric
End: 5

10. Click **OK**.

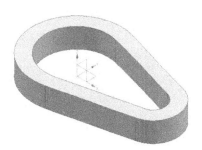

11. On the ribbon, click **Tools > Utilities > Expression** .
12. On the **Expressions** dialog, select **Show > All Expressions**. All the expressions in the file are displayed.
13. Scroll to **p12 (Extrude (2) Start Offset)** in the **Source** column of the expressions sheet.

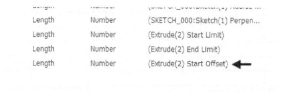

14. Type 3 in the **Formula** box and click **Apply**.

15. Click **OK** to update the model.

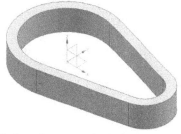

16. Select the **Extrude** feature from the **Part Navigator**.
17. Expand the **Details** section on the **Part Navigator**.
18. Double-click on the **Start Limit** value and type 10. The model is updated, as shown.

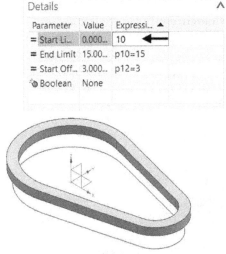

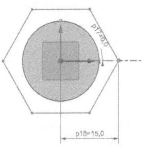

19. Save and close the part file.

TUTORIAL 2

In this tutorial, you will create expressions to drive the parameters of a bolt.

1. Start a new part file.
2. Activate the **Extrude** command and select the YZ plane.
3. Create a circle of 20 mm diameter.

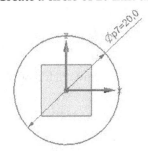

4. Click **Finish**.
5. Extrude the sketch up to 80 mm distance.

6. Activate the **Extrude** command and select the right end face of the cylinder.
7. Create a hexagon, as shown.

8. Click **Finish**.
9. Extrude the sketch up to 10 mm distance.

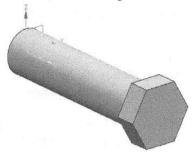

10. On the ribbon, click **Tools > Utilities > Expressions** =.
11. On the **Expressions** dialog, select **Show > All Expressions**.
12. Double click in the **Name** box of (Extrude (1) Diameter Dimension on Arc1).

	↑ Name	Formula
1		
2	p0	0
3	p1	80
4	p7	20
5	p8	0
6	p9	10
7	p15	15

13. Type **Diameter** in the **Name** box and click **Apply**.
14. Double click in the **Formula** box of (Extrude (2) Horizontal Dimension between Line 7 and Line7).

	↑ Name	Formula
1		
2	Diameter	20
3	p0	0
4	p1	80
5	p8	0
6	p9	10
7	p15	15

15. Type **Diameter** in the **Formula** box and click **Apply**.

↑ Name	Formula
1	
2 Diameter	20
3 p0	0
4 p1	80
5 p8	0
6 p9	10
7 p15	Diameter ⬅

16. Double click in the **Formula** box of **(Extrude (2) End Limit)**.
17. Type **0.75*Diameter** in the **Formula** box and click **Apply**.

↑ Name	Formula
1	
2 Diameter	20
3 p0	0
4 p1	80
5 p8	0
6 p9	.75*Diameter
7 p15	Diameter

18. Click **OK** to update the model.

19. Select the **Extrude (1)** feature from the **Part Navigator**.
20. Expand the **Details** section on the **Part Navigator**.
21. Double-click on the **Diameter** value and type 10. The model is updated, as shown.

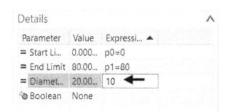

Details ∧

Parameter	Value	Expressi... ▲
= Start Li...	0.000...	p0=0
= End Limit	80.00...	p1=80
= Diamet...	20.00...	10 ⬅
⚙ Boolean	None	

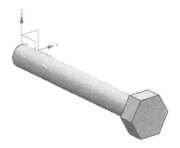

22. On the ribbon, click **Home > Feature > More > Design Feature > Thread** 🔩 .
23. On the **Thread** dialog, set the **Type** to **Detailed**.
24. Select the cylindrical face of the geometry.
25. Click **OK** to create the thread.

26. Press Ctrl+E to open the **Expressions** dialog.
27. On the **Expressions** dialog, select **Show > Feature Expressions**.
28. Select the **Threads** feature from the **Part Navigator** and notice all the expressions related to it.
29. Double click in the **Name** box of **(Threads (3) Major Diameter)**.
30. Type **D** in the **Name** box.
31. Type **Diameter** in the **Formula** box to make the major diameter value equal to the diameter of the cylinder.
32. Double click in the **Name** box of **(Threads (3) Pitch)**.
33. Type **Pitch** in the **Name** box.
34. Double click in the **Name** box of **(Threads (3) Minor Diameter)**.
35. Type **D2** and **D-1.08*Pitch** in the **Name** and **Formula** boxes, respectively.
36. Double click in the **Name** box of **(Threads (3) Length)**.
37. Type **Length** and **2*D** in the **Name** and **Formula** boxes, respectively.
38. Click **Apply** and **OK**. The model is updated, as shown.

39. Select the **Extrude (1)** feature from the **Part Navigator**.
40. Expand the **Details** section on the **Part Navigator**.
41. Double-click on the **Diameter** value and type 20. The model is updated, as shown.

42. Select the **Threads** feature from the **Part Navigator** and change the Pitch value in the **Details** section to 2.5. The pitch and minor diameter of the threads are updated.

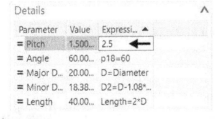

Instead of updating the pitch and diameter values manually, you can use a spreadsheet to change all the values. The following section deals with using a spreadsheet to change the parameters.

Creating Family of Parts

1. On the ribbon, click **Tools > Utilities > Spreadsheet**. The Worksheet in Modeling is opened.
2. In the Worksheet environment, click **ADD-INS > Extract Expr** on the ribbon. The expressions are added to the spreadsheet.
3. Copy the contents of column **B** into column **C** and **D**.

	A	B	C	D
1	**Parameters**			
2	D	20	20	20
3	_D2	17.3	17.3	17.3
4	Diameter	20	20	20
5	Length	40	40	40
6	Pitch	2.5	2.5	2.5
7	_p0	0	0	0
8	_p1	80	80	80
9	_p8	0	0	0
10	_p9	15	15	15
11	_p15	20	20	20
12	_p18	60	60	60

4. Type M20x2.5, M10x1.25, and M6x0.75 in the first rows of columns B, C, and D, respectively.
5. Click in the second row of the column B and change its expression to **=EXPRVAL("Diameter")**

6. Likewise, change the expressions of D values in columns C and D to **=Diameter**.
7. Edit the values of the highlighted rows, as shown.

	A	B	C	D
1	**Paramete**	M20x2.5	M10x1.25	M6x0.75
2	D	20	20	20
3	_D2	17.3	17.3	17.3
4	Diameter	20	10	6
5	Length	40	40	40
6	Pitch	2.5	1.25	0.75
7	_p0	0	0	0
8	_p1	80	60	20
9	_p8	0	0	0
10	_p9	15	15	15
11	_p15	20	20	20
12	_p18	60	60	60

8. Drag the pointer across the A2 and D12 cells.

	A	B	C	D
1	**Paramete**	M20x2.5	M10x1.25	M6x0.75
2	D	20	20	20
3	_D2	17.3	17.3	17.3
4	Diameter	20	10	6
5	Length	40	40	40
6	Pitch	2.5	1.25	0.75
7	_p0	0	0	0
8	_p1	80	60	20
9	_p8	0	0	0
10	_p9	15	15	15
11	_p15	20	20	20
12	_p18	60	60	60

9. Click **ADD-INS > Define Expr Rng** on the ribbon.
10. Click **ADD-INS > Options > NX Preferences**.
11. Uncheck the **Use Fixed Update Range** option and click **OK**.
12. Select the contents of the column D.

	A	B	C	D
1	**Paramete**	M20x2.5	M10x1.25	M6x0.75
2	D	20	20	20
3	_D2	17.3	17.3	17.3
4	Diameter	20	10	6
5	Length	40	40	40
6	Pitch	2.5	1.25	0.75
7	_p0	0	0	0
8	_p1	80	60	20
9	_p8	0	0	0
10	_p9	15	15	15
11	_p15	20	20	20
12	_p18	60	60	60

13. Click **ADD-INS > Update NX Part**.
14. Save and close the spreadsheet. The part is

updated.

15. On the ribbon, click **Tools > Utilities > Spreadsheet** .
16. Select the contents of the column C.
17. Click **ADD-INS > Update NX Part**.
18. Save and close the spreadsheet. The part is updated.

19. On the ribbon, click **Tools > Utilities > Spreadsheet** .
20. Likewise, select the contents of the column **B** and click **Update NX Part**.
21. In the spreadsheet, enter the location and part name (for example:
    C:\Users\Public\Documents\M20x2.5) at the bottom of the B, C, and D columns. The part names should be M20x2.5, M10x1.25, and M6x0.75.

	A	B	C	D
1	**Paramete**	M20x2.5	M10x1.25	M6x0.75
2	D	20	20	20
3	_D2	17.3	17.3	17.3
4	Diameter	20	10	6
5	Length	40	40	40
6	Pitch	2.5	1.25	0.75
7	_p0	0	0	0
8	_p1	80	60	20
9	_p8	0	0	0
10	_p9	15	15	15
11	_p15	20	20	20
12	_p18	60	60	60
13		C:\Users\Public\Documents\M20x2.5	C:\Users\Public\Documents\M10x1.25	C:\Users\Public\Documents\M6x0.75

22. Drag the pointer across the A2 and D13 cells.
23. Click **ADD-INS > Define Fmly Rng** on the ribbon. The selected data will be used to create the part family.
24. Click **ADD-INS > Build Family** on the ribbon.
25. Close the spreadsheet and click **Discard**. The part family is created in the specified folder.

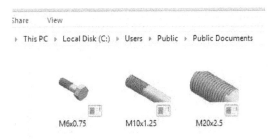

Share   View

▸ This PC ▸ Local Disk (C:) ▸ Users ▸ Public ▸ Public Documents

M6x0.75   M10x1.25   M20x2.5

26. Close the part file.

TUTORIAL 3

1. Download the Tutorial 3 file of the Expressions chapter and open it.

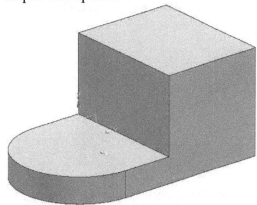

2. On the ribbon, click **Analysis > Measure > More > Simple Length**.
3. Select the edge of the geometry, as shown.

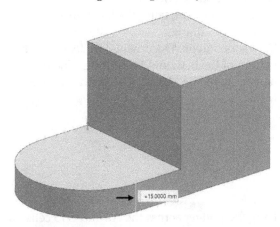

4. Click **OK**; the Length Measurement parameter is displayed in the Part Navigator.

Part Navigator

Name ▲	Up to Date	Comm
⏱ History Mode		
+ 🗗 Model Views		
+ ✔ 🎥 Cameras		
+ 🗂 Measures		
– 🗂 Model History		
☑ 📐 Datum Coordina...	✔	
☑ 🖼 Sketch (1) "SKET...	✔	
☑ 🖼 Extrude (2)	✔	
☑ 🖼 Extrude (3)	✔	
☑)⟩ **Length Measur...**	✔ ←	

5. On the ribbon, click **Tools > Utilities > Expressions**.
6. On the **Expressions** dialog, select **Show > All Expressions**.
7. Double click in the **Name** box of **(Length Measurement (4))**.
8. Type **Thickness** in the **Name** box.

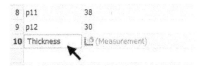

8	p11	38
9	p12	30
10	Thickness	📐 (Measurement)

9. Click **Apply** and **OK**.
10. On the ribbon, click **Home > Feature > Shell**.
11. Select the horizontal face, as shown.

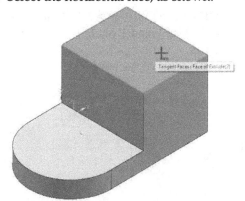

12. Click the down-arrow on Thickness handle and select **Formula**.

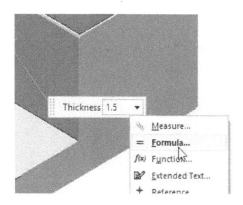

13. On the **Expressions** dialog, type Thickness in the **Formula** box and click **Apply**.

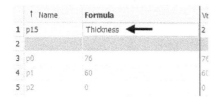

14. Click **OK** on the **Expressions** and **Shell** dialogs.
15. Click on the second extruded feature and select **Show Dimensions**.
16. Double-click on the linear dimension and change its value to 10.

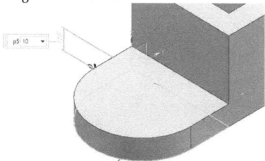

17. Click **OK** on the **Feature Dimension** dialog.
18. Press **F5** on your keyboard.

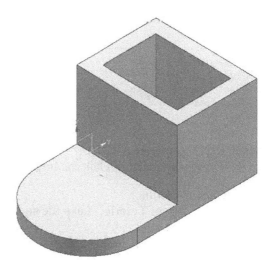

TUTORIAL 4

1. Download the Tutorial 4 file of the Expressions chapter and open it.

2. On the ribbon, click **Tools > Utilities > Expressions**.
3. On the **Expressions** dialog, select **Show > All Expressions**.
4. On the **Expression** dialog, expand the **Import/Export** section, and then click the **Export Expressions** button.
5. Browse to a location to save the file.
6. Set the **Export Options** to **Work Part**.
7. Type **Tutorial_4** in the **File name** box and click **OK**. The expressions of the model are exported to a text file.
8. Open the **Tutorial_4.exp** file in Notepad or any text editor.
9. Modify the expressions in the text file and save it.

```
[mm]Diameter=25
[mm]Extrude1=18
[mm]Extrude2=0.75*Extrude1
[mm]Hole_diameter=Diameter/2
[mm]Parallel_Dimension=90
[mm]Slot_length=Parallel_Dimension/2
[mm]Slot_radius=Hole_diameter/2
[mm]p65=0
[mm]p73=4
```

10. Switch to NX application window.
11. On the **Expressions** dialog, click the **Import Expressions** icon.
12. Go to the location of **Tutorial_4.exp** file and double-click on it.
13. Click **OK** on the **Expressions** dialog.

14. Save and close the file.

Chapter 8: Sheet Metal Modeling

This chapter will show you to:

- Construct Tab feature
- Construct Flange
- Contour Flange
- Closed corners
- Louvers
- Beads
- Drawn Cut-outs
- Gussets
- Flat Pattern

TUTORIAL 1

In this tutorial, you construct the sheet metal model shown in figure.

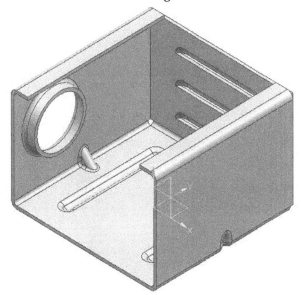

Opening a New Sheet Metal file

1. To open a new sheet metal file, click **Home > New** on the ribbon.
2. On the **New** dialog, click **Sheet Metal**.
3. Click **OK**.

The NX Sheet Metal ribbon appears, as shown below.

Setting the Parameters of the Sheet Metal part

1. To set the parameters, click **File > Preferences > Sheet Metal**

On the **NX Sheet Metal Preferences** dialog, , you can set the preferences of the sheet metal part such as thickness, bend radius, relief depth, width and so on. In this tutorial, you will construct the sheet metal part with the default preferences. Click **OK** on the dialog.

Constructing the Tab Feature

1. To construct the base feature, click **Home >**

    Basic > Tab on the ribbon.
2. Select the XY plane.
3. Construct the sketch, as shown.

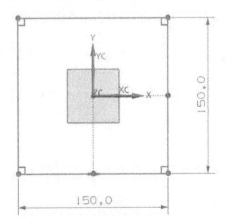

4. Click **Finish**.

5. Click **OK** to construct the tab feature.

Adding a flange

1. To add the flange, click **Home > Bend > Flange**

    on the ribbon.
2. Select the edge on the top face.

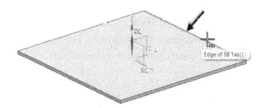

3. On the **Flange** dialog, set **Length** to 100.
4. Click **OK** on the **Flange** dialog to add the flange.

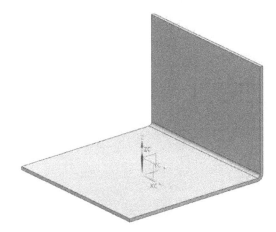

Constructing the Contour Flange

1. To construct the contour flange, click **Home >**

   Bend > Contour Flange on the ribbon.
2. On the **Contour Flange** dialog, click the **Sketch**

   Section icon.
3. On the Top Border Bar, select **Curve Rule >**
   Single Curve.

4. Click on the left edge of the top face at the location, as shown.

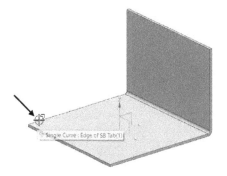

5. On the **Create Sketch** dialog, under the **Plane Location** section, type-in **100** in the % **Arc Length** box.
6. Under the **Plane Orientation** section, click **Reverse Plane Normal**.

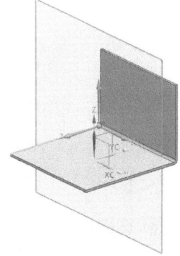

7. Click **OK**.
8. Draw the sketch, as shown.

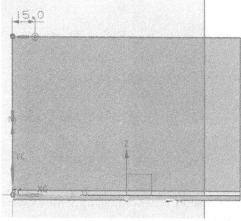

9. Click **Finish**.
10. On the **Contour Flange** dialog, under the **Width**

section, select **Width Option > To End**.

11. Click on the arrow attached to the sketch, if it points in the direction shown below.

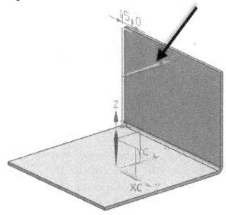

12. Click **OK** to construct the contour flange.

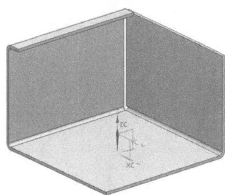

Adding the Closed Corner

1. To add the closed corner, click **Home > Corner > Closed Corner** .

2. Select the two bends forming the corner.

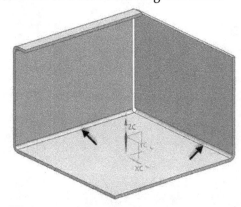

3. On the **Closed Corner** dialog, under the **Corner Properties** section, select **Treatment > Open**.
4. Select **Overlap > Closed**.
5. Click **OK** to add the open corner.

You can also apply corner treatment using the options in the **Treatments** drop-down. The different types of the corner treatments are given next.

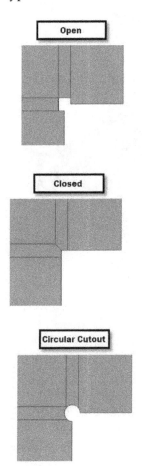

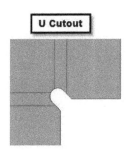

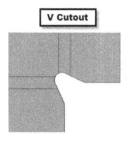

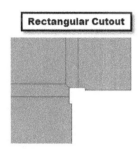

Check the **Miter Corner** option, if you want the result, as shown below.

Adding the Louver

1. To add the louver, click **Home > Punch >**
   Louver on the ribbon.
2. Click on the front face of the flange at the location shown below.

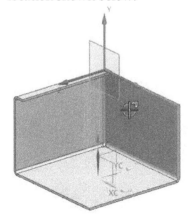

3. Construct the sketch, as shown in figure.

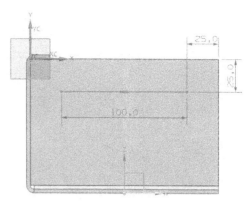

4. Click **Finish**.
5. On the **Louver** dialog, select **Louver Shape > Formed**.
6. Type-in **5** in the **Depth** box and click the **Reverse Direction** icon next to it.
7. Type-in **10** in the **Width** box and click the **Reverse Direction** icon next to it.
8. Click **OK** to add the louver.

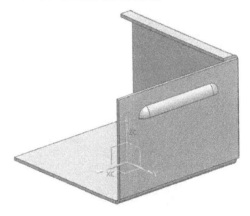

Making the Pattern Along curve

1. On the ribbon, click **Home > Feature > Pattern Feature** .
2. Select the louver feature.
3. On the **Louver** dialog, under **Pattern Definition** section, select **Layout > Along**.
4. Under **Direction 1** section, click **Select Path**.
5. On the Top Border Bar, select **Curve Rule > Single Curve**.

6. Select the vertical edge of the flange feature.

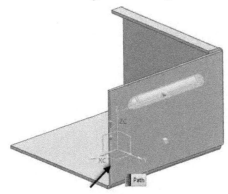

7. Under the **Direction 1** section, select **Spacing > Count and Span**.
8. Set **Count** to 3.
9. Set % **Span By** as 60.
10. Make sure that the arrow points downwards. You can double-click on it to reverse its direction.
11. Click **OK** to construct the pattern along curve.

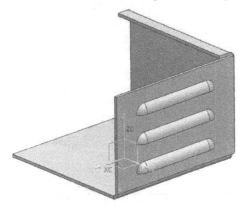

Adding the Bead

1. To add the bead, click **Home > Punch > Bead** on the ribbon.
2. Click on the top face of the tab feature at the location, as shown.

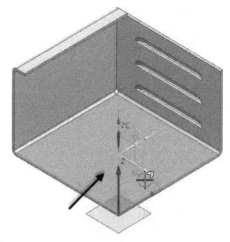

3. Draw a line and dimension it.

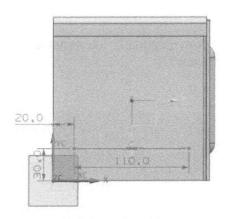

4. Click **Finish** on the ribbon.
5. Under the **Bead Properties** section, select **Cross Section > Circular**.
6. Set **Depth** to 4 and click the **Reverse Direction** icon next to the **Depth** box.
7. Set **Radius** to 4.
8. Select **End Condition > Formed**.
9. Click **OK** to add the bead.

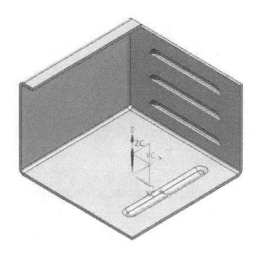

Adding the Drawn Cutout

1. To add the drawn cutout, click **Home > Punch >**

    Drawn Cutout on the ribbon.
2. Click on the face of the contour flange at the location shown below.

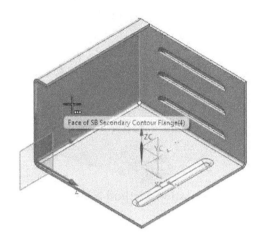

3. Draw a circle and dimension it.

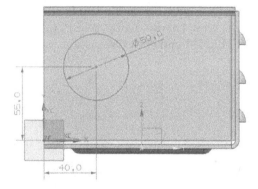

4. Click **Finish** on the ribbon.
5. Set **Depth** to 10.
6. Set **Side Angle** to 5.
7. Select **Side Walls > Material Outside**.
8. Expand the **Rounding** section and uncheck the **Round Section Corners** option.
9. Set **Die Radius** to 3.
10. Click **OK** to add the drawn cutout.

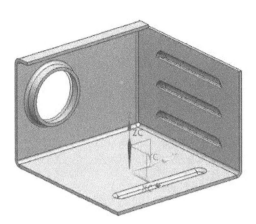

Adding Gussets

1. To add gussets, click **Home > Punch > Gusset** on the ribbon.
2. Click on the bend face of the contour flange.
3. On the **Gusset** dialog, select **Type > Automatic Profile**.
4. Under the **Location** section, select **YC** from the drop-down.

5. Under the **Shape** section, set **Depth** to 12.
6. Select **Form > Round**.
7. Set **Width** to 10.
8. Set **Side Angle** to 2.
9. Set **Die Radius** to 2.
10. Click **OK** to add gussets.

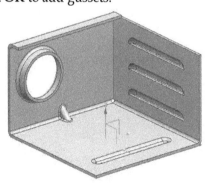

Constructing the Mirror Feature

1. To construct the mirror feature, click **Home >**

    Feature > More > Mirror Feature on the ribbon.
2. Under the **Part Navigator**, press the Ctrl key, and then select the contour flange, closed corner, bead feature, and gusset.

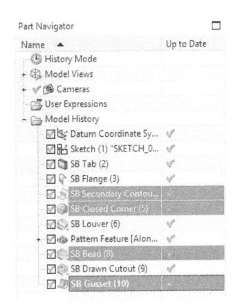

3. Under the **Mirror Plane** section, click **Select Plane**.
4. Select the YZ plane.

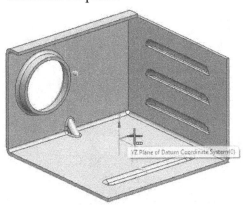

5. Click **OK** to construct the mirror feature.

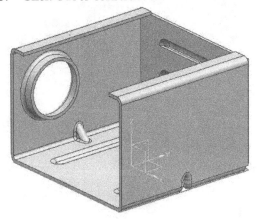

Making the Flat Pattern

1. To make the flat pattern, click **Home > Flat**

Pattern > Flat Pattern on the ribbon.
2. Click on the top face of the tab feature.

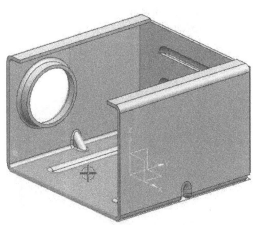

3. Select **Orientation Method > Default** from the **Orientation** section.
4. Click **OK** to make the flat pattern.
5. On the **Sheet Metal** message, click **OK**.
6. To view the flat pattern, click **Menu > View > Layout > > New** on the Top Border Bar.
7. On the **New Layout** dialog, select FLAT-PATTERN#1 and click **OK**.

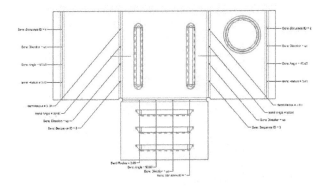

8. To view the 3D model, click Menu > **View > Layout > New** on the Top Border Bar.
9. On the **New Layout** dialog, select **Isometric** and click **OK**.

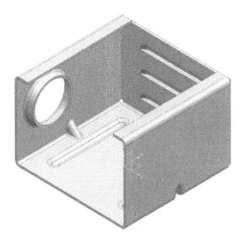

10. Save and close the file.

TUTORIAL 2

In this tutorial, you will apply corner relief to 3 Bend corner.

1. Download and open the Tutorial_2 part file from the companion website.
2. On the CommandManager, click **Home > Corners > Corners** gallery > **Three Bend Corner**.

3. Select the two bend faces adjacent to the corner, as shown.

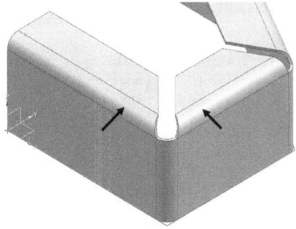

4. On the Three Bend Corner dialog, under the **Corner Properties** section, select **Treatment > Circular Cutout** .
5. Uncheck the **Miter Corner** option.

6. Type 10 in the **Diameter** box available in the **Relief Properties** section.
7. Click **OK** on the **Three Bend Corner** dialog.

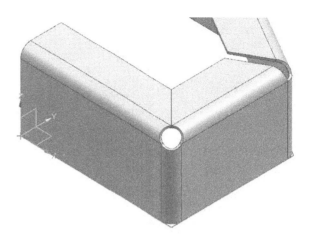

TUTORIAL 3

In this tutorial, you will create a multibody sheet metal part.

1. Start a new NX Sheet metal part file.
2. On the ribbon, click **File** tab > **Preferences > Sheet Metal**.
3. Leave the default settings on the **Sheet Metal Preferences** dialog, and then click **OK**.

4. On the ribbon, click **Home > Basic > Tab** .
5. Click on the XY plane.
6. Create a rectangle and add dimensions to it, as shown.

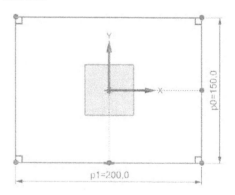

7. Click **OK**.

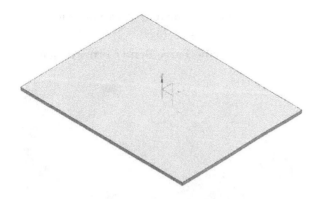

8. Create flanges of 40 mm length, as shown.

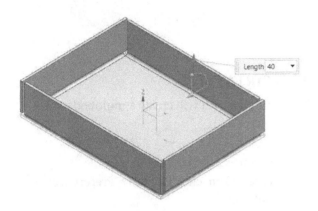

9. On the ribbon, click **Home** tab > **Corner** group > **Closed Corner** .

10. On the **Closed Corner** dialog, select **Type > Close and Relief**.

11. Select the two bends forming a corner, as shown.

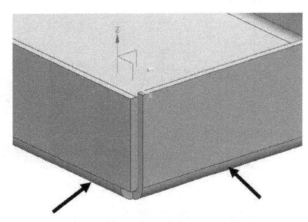

12. On the **Closed Corner** dialog, under the **Corner Properties** section, select **Treatment > Open** .

13. Select **Overlap > Closed**.

14. Click **OK** to close the corner and apply the relief.

15. Likewise, apply the corner relief to the remaining corners.

16. Activate the **Tab** command.

17. On the **Tab** dialog, select **Type > Base**.

18. Click on the outer face of the flange.

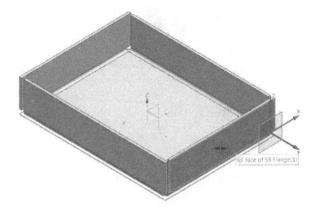

19. Create a rectangle, as shown.

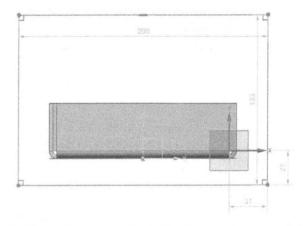

20. Press Esc to deactivate the **Rectangle** command.

21. Select the bottom edge and the bottom horizontal line of the rectangle.

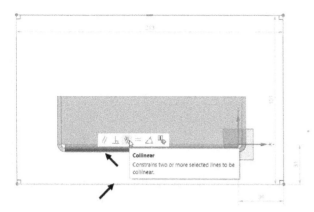

22. Click the **Collinear** icon the Context toolbar; the line is made collinear with the horizontal edge of the bend.
23. Likewise, make the vertical line collinear with the side edges, as shown.

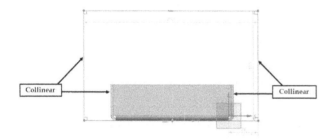

24. Add dimension to the vertical line, as shown.

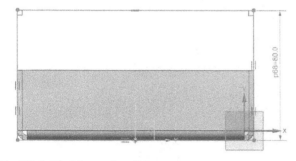

25. Click **Finish** on the ribbon.
26. Check the **Reverse Direction** option.
27. Click **OK**; a separate tab feature is created.

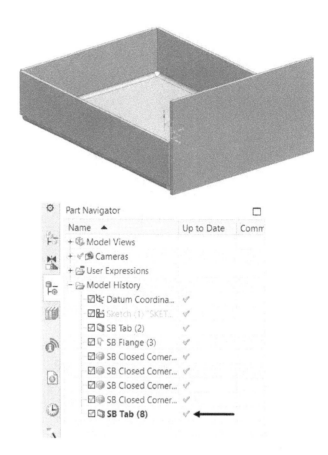

28. Click on the Z-axis of the triad located at the bottom left corner of the graphics window.
29. Type **180** in the **Angle** box, and then press; the orientation of the model is changed.

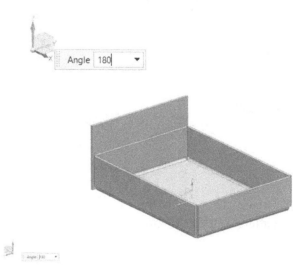

30. Activate the **Tab** tool.
31. On the Tab dialog, select **Type > Base**.
32. Click on the edge flange face.

33. Activate the **Rectangle** tool.
34. Click on the first and second corners, as shown.

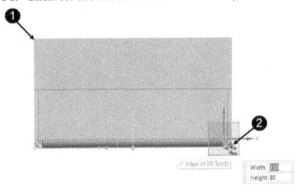

35. Click **Finish** on the ribbon.
36. Click **OK**.

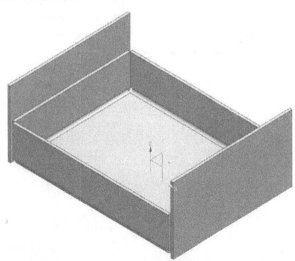

37. On the ribbon, click **Home > Bend > More gallery > Bridge Blend**.

38. On the **Bridge Blend** dialog, select **Type > Z or U Transition**.
39. On the **Bridge Blend** dialog, under the Width section, select **Width Option > Full Both Edges**.
40. Click on the edge of the third sheet metal body, as shown.

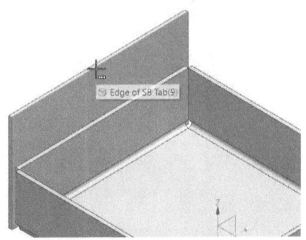

41. Select the edge of the second tab feature, as shown.

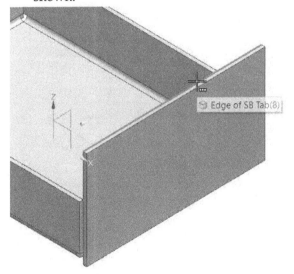

42. Click **OK** on the dialog.

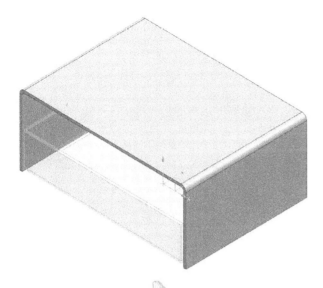

45. On the **Flange** dialog, select **Width Option > Full**.
46. Select **Length Reference > Web**.
47. Select **Inset > Material Outside**.
48. Type **80** in the **Length** box.
49. Select the edge on the other side of the bridge
50. Expand the **Relief** section, and then select **Corner Relief > Bend/Face Chain**.
51. Click **OK**.

43. Activate the **Flange** tool.
44. Click on the edge of the previously created edge flange, as shown.

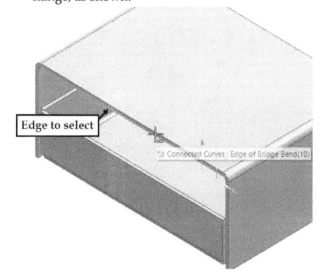

Chapter 9: Top-Down Assembly

In this chapter, you will learn to

- Create a top-down assembly
- Insert fasteners
- Create Sequences
- Create Deformable Parts and assemble them

TUTORIAL 1

In this tutorial, you will create the model shown in figure. You use top-down assembly approach to create this model.

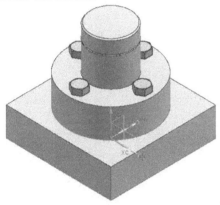

Creating a New Assembly File
1. Click the **New** icon on the Quick Access Toolbar, select the **Assembly** template, and click the **Browse** icon located next to the **Name** box.
2. Create a new folder and open it.
3. Type Tutorial 1 in the **File Name** box.
4. Click **OK** twice.
5. Click **Cancel** on the **Add Component** dialog.

Creating a component in the Assembly
In a top-down assembly approach, you create components of an assembly directly in the assembly by using the **Create New** tool.

1. On the ribbon, click **Assemblies > Component > Create New** .
2. Select the **Model** template, type **Base** in the **Name** box, and click **OK**.
3. Click **OK** on the **Create New Component**

dialog.
4. Click the **Assembly Navigator** tab on the **Resource Bar**.

5. Double-click on the **Base** component. The part mode is activated.

6. Click **Home > Sketch** on the ribbon.
7. Select **Sketch Type > On Plane** from the Create Sketch dialog.
8. Select the XY plane from Datum Coordinate System and click **OK**.
9. Create the sketch as shown below.

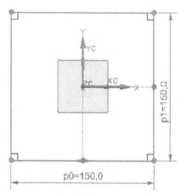

2. Click **Finish Sketch**.
3. Click **Home > Feature > Extrude** on the Ribbon and extrude the sketch up to 40 mm.

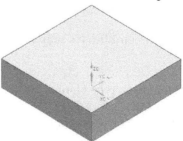

4. Create a cylinder of 50 mm diameter and 95 mm length on the top face.

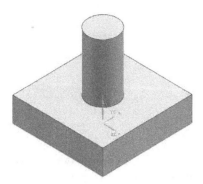

5. On the ribbon, click **Home > Feature > Hole**.
6. On the **Hole** dialog, select **Type > General Hole**.
7. In the **Form and Dimensions** section, set the parameters, as shown.

   Form: Counterbored
   C-Bore Diameter: 30
   C-Bore Depth: 12
   Diameter: 25
   Depth Limit: Through Body

8. Select the center point of the top circular edge.
9. Click **OK**.

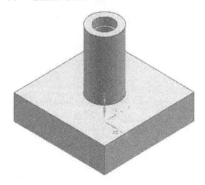

10. On the ribbon, click the **Assemblies > Context Control > Work on Assembly**.

Creating the Second Component of the Assembly

1. On the ribbon, click **Assemblies > Component > Create New**.
2. Select the **Model** template, type **Flange** in the **Name** box, and click **OK**.
3. Click **OK** on the **Create New Component** dialog.
4. In the Assembly Navigator, double-click on the **Flange** to activate the **Work part** mode.
5. Click **Home > Direct Sketch > Sketch** on the Ribbon.

6. On the Top Border Bar, set the **Selection Scope** to **Entire Assembly**.
7. Select top face of the Base.

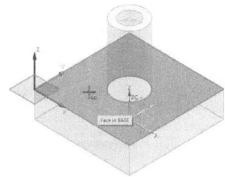

8. Click **OK**.
9. On the ribbon, click **Home > Direct Sketch > Project Curve** .
10. On the Top Border Bar, click the **Create Interpart Link** icon.
11. Select the circular edge of the Base.

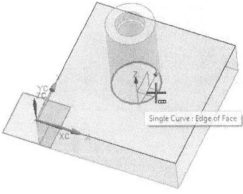

12. Click **OK**.
13. Click **Yes**.
14. Draw a circle of 120 mm diameter.

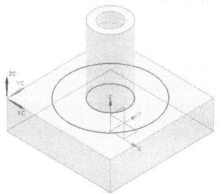

15. Click **Finish Sketch**.
16. Activate the **Extrude** tool and extrude the sketch up to 40.

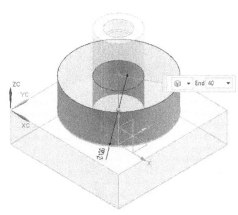

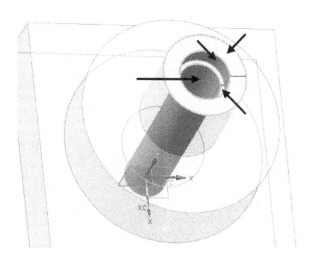

17. Click the **Part Navigator** tab on the Resource Bar and notice the **Linked Component Curve(1)**.

18. In the **Assembly Navigator**, double-click on **Tutorial 1** to switch to the assembly mode.

Creating the third Component of the Assembly

1. On the ribbon, click **Assemblies > Component > Create New** .
2. Select the **Model** template, type **Head Screw** in the **Name** box, and click **OK** twice.
3. In the **Assembly Navigator**, right click on **Head Screw** and select **Make Work Part**.
4. Start a sketch on the XZ Plane.
5. On the ribbon, click **Home > Direct Sketch > More Curve > Intersection Curve** .
6. On the Top Border Bar, make sure that the **Create Interpart Link** icon is active.
7. On the Top Border Bar, set the **Selection Scope** to **Entire Assembly**.
8. Rotate the view and select the faces, as shown.

9. Click **OK**.
10. Right click and select **Orient View to Sketch**.
11. Draw the other lines, as shown.

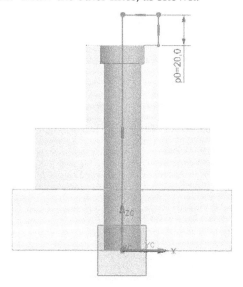

12. Click **Finish Sketch**.
13. Activate the **Revolve** tool and revolve the sketch.

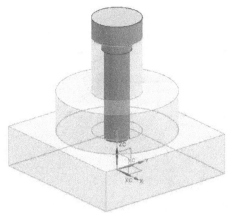

14. Activate the **Chamfer** tool and chamfer the edges, as shown in figure.

15. Activate the **Edge Blend** tool and round the edges, as shown in figure.

16. On the ribbon, click **Home > Assemblies > Work on Assembly**.

Editing the Linked Parts

1. Right click on the Base and select the **Make Work Part**.
2. Select the face of the Base, as shown.
3. Click **Show Dimensions** on the Shortcuts toolbar.

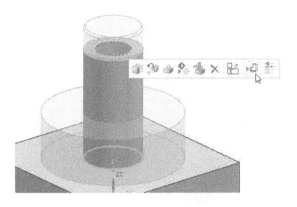

4. Change the Diameter dimension to 60 and Linear dimension to 80.
5. Activate the Assembly mode and notice that the linked parts are also modified.

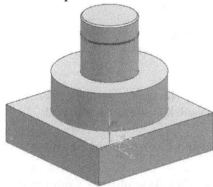

17. On the ribbon, click **Assemblies > General > Interpart Link Browser** . The **Interpart Links Browser** dialog has two sections: **Parts** and **Interpart Links in Selected Parts**.

You can edit a link by selecting it and clicking the **Edit** icon. You can also use the **Break Link** icon to remove the link.

18. Close the **Interpart Link Browser** dialog.

Creating Hole Series

A hole series is created through different parts of the assembly.

1. On the ribbon, click **Home > Feature > Hole**.
2. Select the top face of the Flange.
3. Select **Type > Hole Series**.
4. Position the hole using the Reference line, as shown.

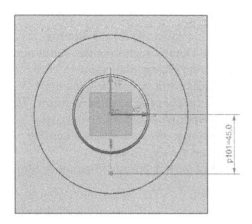

5. Click **Finish**.

Under the **Specification** section, notice the three tabs: Start, Middle, and End. They are used to set the hole parameters for the three bodies through which the hole passes. For this example, you are required to only set the Start and End parameters.

6. Under the **Specification** section of the **Hole** dialog, click the **Start** tab and set the parameters, as shown.

   Form: Simple
   Screw Type: General Screw Clearance
   Screw Size: M12
   Fit: Normal (H13)

7. Click the **End** tab and set the parameters, as shown.

   Form: Threaded
   Depth Type: Full
   Handedness: Right Handed
   Depth Limit: Through Body

8. Click **OK**.
9. Likewise, create three more series holes.

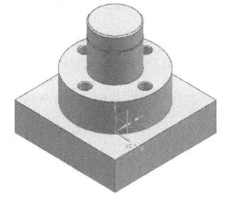

Adding Fasteners to the assembly

1. On the ribbon, click **Tools > Reuse Library > Fastener Assembly** .
2. On the **Fastener Assembly** dialog, select **Type >Hole**.
3. Select anyone of the holes.
4. Click the **Add Fastener Assembly** icon.
5. Set the **Configuration Name** to **AM-Hex Bolt/Stacks**.
6. Under the **Fastener Configuration** section, click the **Remove** icon next to **Plain Washer, Regular, AM** under **Top Stacks**.

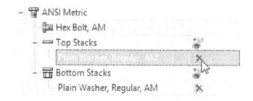

7. Click **OK**.
8. On the **Configuration** section, click the **Properties** icon next to **Hex Bolt, AM**.
9. On the **Edit Reusable Component** dialog, set the **(L) Length** value to **100**. Also, notice the parameters of the hex bolt in the **Details** section. They are read only.
10. Click **OK**.
11. Expand the **Settings** section and check the **Create Constraints Automatically** option.
12. In the **Configuration** section, right click on **AM-Hex Bolt Stacks** and select **Save Configuration**.

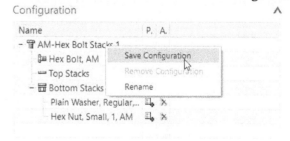

13. Type **Fastener 1** in the **Name** box and click **OK**.
14. Click **OK** to add the fastener assembly.

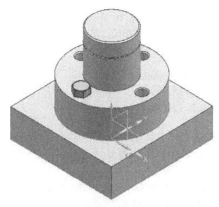

15. Likewise, create add fasteners to other holes.

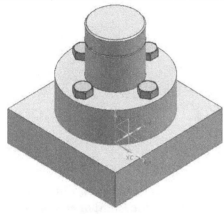

16. Save the assembly and all its parts.

TUTORIAL 2

In this tutorial, you create a sequence of the assembly.

1. Download the Tutorial 2 files of Chapter 9.
2. Open the Tutorial_2 assembly file.

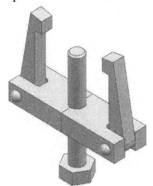

3. On the ribbon, click **Assemblies > General > Sequence** .
4. On the ribbon, click **Home > Assembly Sequence > New** .

5. On the Resource Bar, click the **Sequence Navigator** tab and select **Sequence_1**.
6. Expand the **Details** section of the **Sequence Navigator** and change the **Name** to Hub Puller.
7. In the **Details** section, double-click in the **Value** column of the **Display Split Screen** row. The graphics window is split into two parts.
8. Drag a selection box around the assembly displayed on the right side.
9. On the ribbon, click **Home > Sequence Steps > Disassemble** . The disassemble events are displayed under the **Preassembled** folder of the **Sequence Navigator**.

Part_3	
Part_3	10
Part_1	20
Part_2	30
Part_3	40
Part_4	50
Part_4	60

10. On the ribbon, click **Home > Playback > Play Backwards** . Notice that the parts are assembled back in a random sequence.
11. In the Sequence Navigator, press the Ctrl key and select all the events under the **Preassembled** folder.
12. Right click and select **Delete**.
13. Expand the **Preassembled** folder.
14. Press and hold the Ctrl key, and then select the two instances of Part_4.

15. Click the **Disassemble Together** icon on the ribbon.
16. Select the **Sequence Group 1** from the Sequence Navigator and change its Name to Pins.
17. Select the Part_3 and click the **Disassemble** icon on the ribbon.
18. Likewise, disassemble the other instance of Part_3, Part_2, and Parte_1.
19. On the ribbon, click the **Record Camera Position** icon.
20. In the Sequence Navigator, select the **Pins** event and change the **Total Duration** value in the **Details** section to 2.
21. Likewise, change the **Total Duration** values of other events to 2.
22. On the **Playback** group of ribbon, change the **Playback Speed** to **10**.
23. On the **Playback** group, click the **Export to Movie** icon.

24. Type **Hub Puller assembly** in the **File name** box and **OK**. The movie of the assembly sequence is recorded.
25. Click **OK** on the **Export to Movie** message.
26. Click **Finish**.
27. Open and play the video.
28. Close all the parts.

TUTORIAL 3

In this tutorial, you create a deformable part and add it to an assembly.

Creating the Deformable Part

1. Download the Tutorial 3 files of Chapter 9.
2. Open the Deformable_part.prt file.

3. On the Top Border Bar, click **Menu > Tools > Define Deformable Part**.
4. Leave the default **Name** value and click **Next**.
5. Press and hold the Ctrl key, and then select all the features from the **Features in Part** list.
6. Click **Add Feature** ➡.
7. Click **Next**.
8. Select **Pitch = 15** from the **Available Expressions** list and click **Add Expression** ➡.
9. Type **Pitch** in the box below the **Deformable Input Expressions** list.
10. Set the **Expression Rules** to **By Number Range**.
11. Type 8 and 15 in the **Minimum** and **Maximum** boxes.
12. Click **Next** and **Finish**.
13. Save and close the file.

Adding the Deformable part to an Assembly

1. Open the Tutorial 3 assembly file.

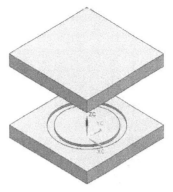

2. On the ribbon, click **Assemblies > Component > Add**.
3. Click the **Open** icon on the **Add Component** dialog.
4. Go to the location of the Deformable_part.prt file and double-click on it.
5. Set **Placement** to **Constrain**.
6. Under the **Settings** section, select **Reference set > Entire Part**.
7. Under the Placement section, select **Constraint Type > Touch Align**.
8. Under the **Geometry to Constrain** section, select **Orientation > Touch**.
9. Select the bottom flat face of the deformable part.

10. Select the flat face of the plate, as shown.

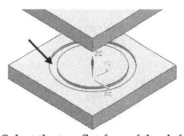

11. Select the top flat face of the deformable part.

12. Select the flat face of the upper plate, as shown.

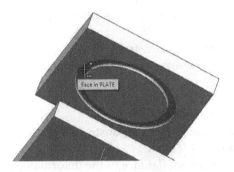

13. Under the **Geometry to Constrain** section, select **Orientation > Infer Center/Axis**.
14. Select the Z-axis of the deformable part.

15. Select the select circular edge of the plate.

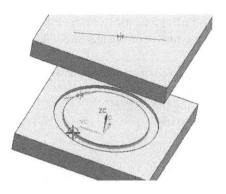

16. Click **OK**.
17. Change the **Pitch** value on the **Deformable_part** dialog to 10 by dragging the slider.
18. Click **OK**.

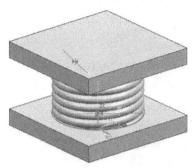

19. On the Resource Bar, click the Part Navigator tab.
20. Right click on the **Deformable_part** feature and select **Edit Parameters**.
21. Drag the slider to change the pitch value to **15**.
22. Click **OK**. The assembly is updated.

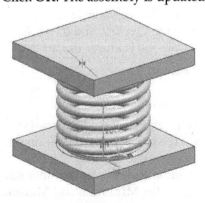

23. Save and close the files.

Chapter 10: Dimensions and Annotations

In this chapter, you will learn to

- Create Centerlines and Center Marks
- Edit Hatch Pattern
- Apply Dimensions
- Place Datum Feature
- Place Feature control frame
- Place Surface Finish symbol

TUTORIAL 1

In this tutorial, you create the drawing shown below.

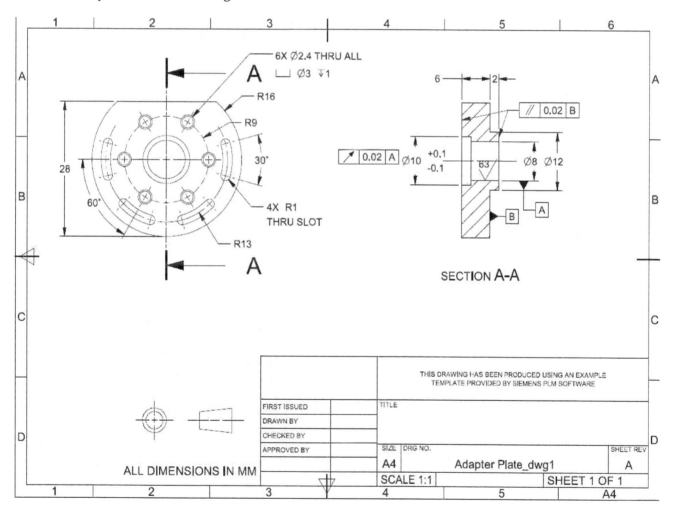

1. Download the Adapter Plate file of Chapter 10.
2. Start NX 12 and click the **New** icon on the ribbon.
3. Click the **Drawing** tab, select **Relationship > Reference Existing Part**.
4. Select the A4 template.
5. Click the **Browse** icon under the **Part to create a drawing of**.
6. Click **Open** on the **Select master part**.
7. Go to the location of the Adapter Plate file and double-click on it.
8. Click **OK** on the **Select master part** and **New** dialogs.
9. Type your own values on the **Populate Title Block** dialog and click **Close**.

Creating a View with Center Marks

1. On the ribbon, click **Home > View > View Creation Wizard** .
2. Click the **Reset** button on the **View Creation Wizard**.

3. Click the **Next** button on the **View Creation Wizard**.
4. On the **Options** page, leave the **Show Centerlines** option selected.
5. Click **Next**.
6. Select **Front** view and click **Finish**.

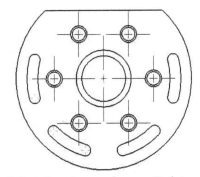

7. Select the view and press **Delete**.
8. On the ribbon, click **Home > View > Base View**.
9. On the **Base View** dialog, expand the **Settings** section and click the **Settings** icon.
10. On the **Settings** dialog, click **General** from the tree.
11. On the **General** page, uncheck the **Create with**

Centerlines option and click **OK**.
12. Select **Model View to Use > Front**.
13. Set the **Scale** value to **2:1**.
14. Click on the drawing page, as shown.

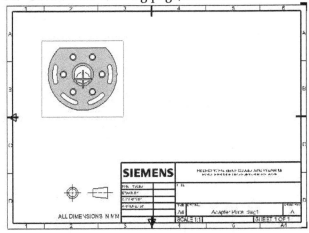

15. Close the **Projected View** dialog.
16. Click **Home > View > Section View** on the Ribbon.
17. Select the center point of the front view.
18. Place the section view on the right side.
19. Click **Close**.

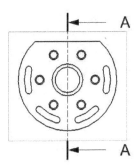

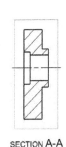

SECTION A-A

Creating Centerlines and Center Marks

1. Click **Home > Annotation > Center Mark > Bolt Circle Centerline** on the Ribbon.
2. On the **Bolt Circle Centerline** dialog, select **Type > Through 3 or More Points**.
3. Leave the **Full Circle** option checked.
4. Select the counterbore hole pattern.
5. Drag the arrow that appears on the centerline to change its Extension length.

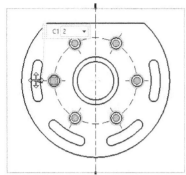

6. Click **OK**.
7. Click **Home > Annotation > Centerline drop-down> Circular Centerline** on the Ribbon.
8. On the **Circular Centerline** dialog, uncheck the **Full Circle** option.
9. Select the center points of the arcs, as shown.

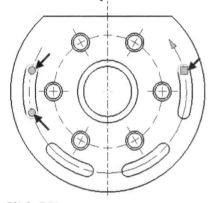

10. Click **OK**.
11. Click **Home > Annotation > Centerline drop-down> 2D Centerline** on the Ribbon.
12. On the **2D Centerline** dialog, select **Type > By Points**.
13. On the Top Border Bar, activate the **Control Point** and **Intersection** icons, and deactivate the **Arc Center** icon.

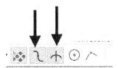

14. Select the points on the slot, as shown.

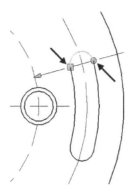

15. Drag the arrow to reduce the length of the centerline.

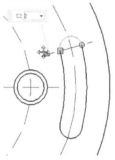

16. Click **Apply**.
17. Likewise, create centerlines on other slots, as shown.

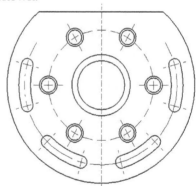

18. Click **Home > Annotation > Centerline drop-down> Automatic Centerline** on the Ribbon.
19. Select the front view and click **OK**; the remaining centerlines are created, as shown.

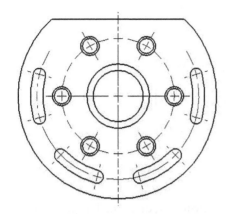

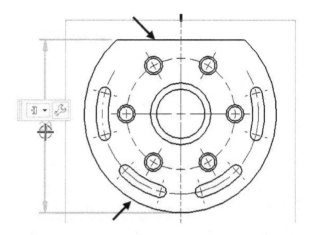

Editing the Hatch Pattern

1. Double-click on the hatch pattern of the section view. The **Crosshatch** dialog appears.
2. On the **Crosshatch** dialog, expand the **Settings** section and notice options to modify the hatch pattern.

You can select the required hatch pattern from the **Pattern** drop-down. You can adjust the distance, angle, color, width, boundary curve tolerance. You can also select a different set of hatch patterns from the **Crosshatch Definition** drop-down.

3. Click **OK**.

Applying Dimensions

1. On the Top Border Bar, click **Menu > Tools > Drafting Standard**.
2. On the **Load Drafting Standard** dialog, select **Standard > ASME**.
3. Click **OK**.
4. Click **Home > Dimension > Rapid** on the Ribbon.
5. On the **Rapid Dimension** dialog, under the **Measurement** section, select **Method > Vertical**.
6. Select the horizontal edge and the outer arc of the front view.
7. Move the pointer toward left and click.

8. On the **Rapid Dimension** dialog, select **Method > Radial**.
9. Create radial dimensions by selecting the circular centerlines, outer arc, and slot arc.

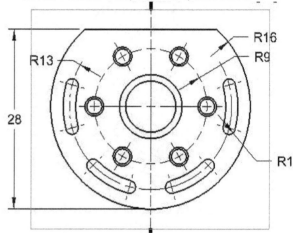

10. On the **Rapid Dimension** dialog, select **Method > Diametral**.
11. Select the counterbore hole and position the diameter dimension, as shown.

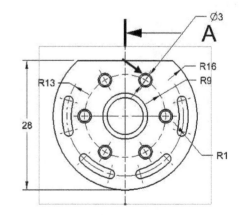

12. On the **Rapid Dimension** dialog, select **Method > Angular**.

13. Select the 2D centerlines of the slot and position the angular dimension, as shown.

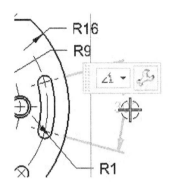

14. Likewise, create another angular dimension, as shown.

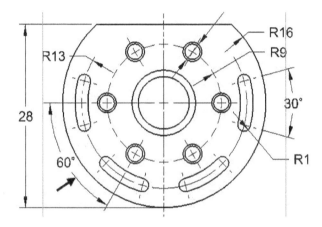

15. On the **Rapid Dimension** dialog, under the **Measurement** section, select **Method > Cylindrical**.
16. Zoom to the section view and select the end points, as shown.
17. Move the pointer right and position the dimension.

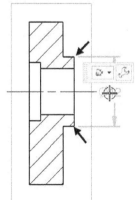

18. Select the horizontal edges of the hole and position the dimension, as shown.

19. Create another cylindrical dimension for the counterbore hole.
20. Click **Home > Dimension > Linear Dimension** on the Ribbon.
21. Select the vertices of the section view, as shown.

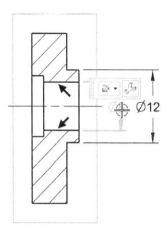

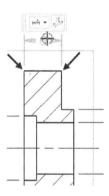

22. Move the pointer up and place the dimension.
23. On the **Linear Dimension** dialog, expand the **Dimension Set** section and select **Method > Chain**.
24. Select the vertex of the section view, as shown.

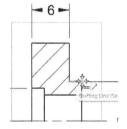

25. Click **Close** on the dialog.
26. Drag the dimension 6 toward left.

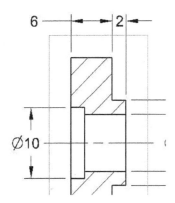

Attach Text to Dimensions

1. Zoom to the front view and double-click on diameter 3.
2. Click the **Arrows Out Diameter** icon on the palette.
3. Click the **Edit Appended Text** icon.
4. On the **Appended Text** dialog, select **Text Location > Above**.
5. Type the text in the box available on the dialog. Also, use the diameter symbol available in the **Symbol** section.

6. Select **Text Location > Before**.
7. Click the **Insert Counterbore** icon in the **Symbols** section.
8. Select **Text Location > After**.
9. Click the **Insert Depth** icon in the **Symbols** section and type 1.
10. Click **Close** on the dialog.
11. Drag the dimension, as shown.

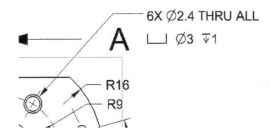

12. Likewise, attach text to the radius dimension of

the slot.

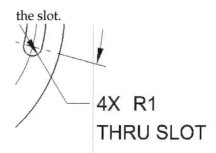

13. Double-click on the radius dimension.
14. Click on the square dot attached to the arrow.
15. Select the **Out** option from the handle.

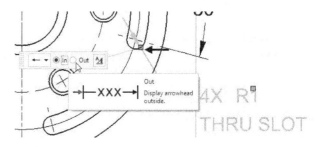

16. Likewise, change the arrow direction of the other radial dimensions.

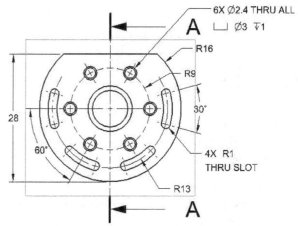

17. Double the counterbore dimension of the section.
18. On the palette, select **Bilateral Tolerance** from the Tolerance drop-down.
19. Type +0.1 and -0.1 in the tolerances boxes.

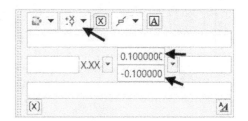

20. Click **Close**.

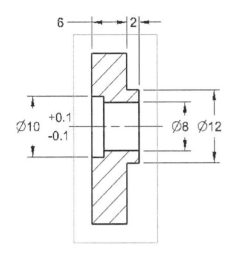

Placing the Datum Feature Symbol

1. Click **Home > Annotation > Datum Feature Symbol** on the Ribbon.

2. On the **Datum Feature Symbol** dialog, expand the **Leader** section and click **Select Terminating Object**.

3. Select the extension line of the dimension, as shown below.

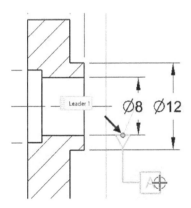

4. Move the cursor downward and click.

5. On the **Datum Feature Symbol** dialog, type **B** in the **Letter** box under the **Datum Identifier** section.

6. Click **Select Terminating Object** and select the vertical edge of the section view, as shown.

7. Move the pointer towards right and click.

8. Click **Close**.

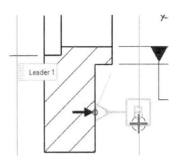

Placing the Feature Control Frame

1. Click **Home > Annotation > Feature Control Frame** on the Ribbon.

2. On the dialog, select **Circular Runout** from the **Characteristic** drop-down.

3. Type-in **0.02** in the **Tolerance** box.

4. Select **A** from the **Primary Datum Reference** drop-down.

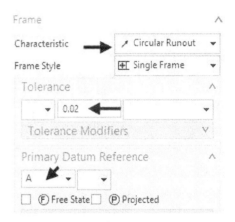

5. Place the pointer on the counterbore diameter dimension.

6. Click when a dashed rectangle appears.

7. On the dialog, select **Parallelism** from the **Characteristic** drop-down.

8. Type-in **0.02** in the **Tolerance** box.

9. Select **B** from the **Primary Datum Reference** drop-down.

9. Expand the Leader section, and click **Select Terminating Object**.

10. Select an edge parallel to the Datum B.

11. Click **Select Terminating Object** and select another edge which is parallel to the Datum B.

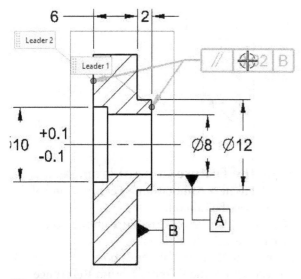

10. Click **Close**.

Placing the Surface Texture Symbols

1. Click **Home > Annotation > Surface Finish Symbol** √ on the Ribbon.
2. Set the **Roughness (a)** value to 63 on the dialog.
3. Click on the inner cylindrical face of the hole, as shown below.

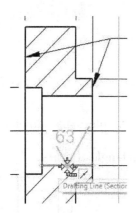

4. Click **Close**.
5. Save and close the file.

Chapter 11: Simulation Hands on Tutorial

TUTORIAL 1

In this tutorial, you perform Finite Element Analysis on a part.

1. Download the Tutorial 1-part file of Chapter 11, and open it.

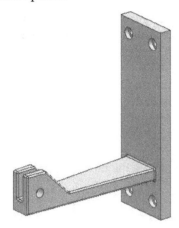

2. On the ribbon, click **Application > Simulation > Pre/Post** .
3. On the Simulation Navigator, select Tutorial 1.prt.
4. On the ribbon, click **Home > Context > New FEM and Simulation** .
5. Leave the **Create Idealized Part** option checked. Notice the three file types (FEM, Simulation, and Idealized): **Tutorial 1_fem1.fem, Tutorial 1_sim1.sim, Tutorial 1_fem1_i.prt** displayed on the dialog. Note that three files are created in addition to the main part file.
6. Under the **Solver Environment** section select **Solver > NX Nastran**.
7. Select **Analysis Type > Structural**.
8. Click **OK**.
9. Leave the default options on the **Solution** dialog and click **OK**.

On the **Simulation Navigator**, notice the **Status** of the Simulation and FEM files.

10. Expand the **Simulation File View** section, right click on **Tutorial 1_sim1**, and click **Save**. The simulation tools are displayed on the ribbon.

Preparing the Idealized Part

1. Hide the **Simulation File View** section.
2. On the **Simulation Navigator**, expand the Tutorial 1_fem1.fem node, right click on Tutorial 1_fem1_i.prt, and click **Make Displayed Part** (or) click **View > Window > Tutorial 1_fem1_i.prt** on the ribbon.
3. Click **OK** on the **Idealized Part Warning** message box. Notice the Status of the idealized part.

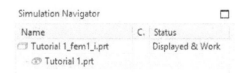

4. On the ribbon, click **Home > Start > Promote** .
5. Select the geometry from the graphics window and click **OK**. The program establishes an associative link between the idealized part and the main part file.

Now, you need to prepare the idealized part by removing some features such as holes and blends.

6. On the ribbon, click **Home > Synchronous Modeling > Delete Face** .
7. On the **Delete Face** dialog, select **Type > Hole**, and uncheck the **Select Holes by Size** option.
8. Select the cylindrical face of the counterbore, as shown.

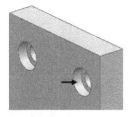

9. Select the other counterbore holes, as shown.

10. Click **Apply** to delete the counterbore holes.
11. On the **Delete Face** dialog, select **Type > Blend**.
12. Select the edge blends of the geometry and click **OK**.

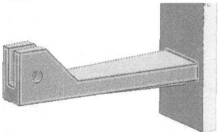

13. Click **Save** on the **Quick Access Toolbar**. Now, you need to switch to FEM file.

Meshing the FEM file

1. On the ribbon, click **Home > Context > Change Displayed Part**.
2. Select **Tutorial 1 _fem1.fem** and click **OK**. The **Information** window appears showing the **CAE Polygon Update Log**.
3. Close the **Information** window. Also, notice the **Status** of the **Tutorial 1 _ fem1.fem** file on the Simulation Navigator.
4. On the ribbon, click **Home > Properties > Mesh Collector**.
5. On the **Mesh Collector** dialog, select **Element**

Family > 3D.

6. Click the **Create Physical Properties** icon.
7. On the **PSOLID** dialog, type Cantilever in the **Name** box.
8. Click the **Choose Material** icon.
9. On the **Material List** dialog, select **Steel** from the **Material** section and click **OK**.
10. Click **OK** on the **PSOLID** and **Mesh Collector** dialogs.
11. On the ribbon, click **Home > Mesh > 3D Tetrahedral**.
12. Select the geometry from the graphics window.
13. On the **3D Tetrahedral Mesh** dialog, select **Type > CTETRA (10)**. You can also set the element type to **CTETRA (4).**
14. Set the **Element Size** to **3**.
15. Expand the **Destination Collector** section, uncheck the **Automatic Creation** option, and make sure that the **Mesh Collector** is set to **Solid (1)**.
16. Click **OK** to generate the mesh.

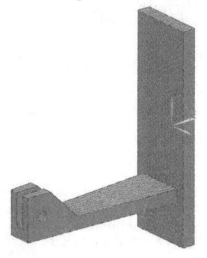

You can edit or remove the mesh from the Simulation Navigator.

17. Expand the **3D Collectors** node in the **Simulation Navigator** and notice the mesh properties.

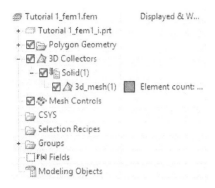

18. Click **Save** on the Quick Access Toolbar.

Applying Loads and Constraints to the Simulation file

1. On the ribbon, click **Home > Context > Change Displayed Part** .
2. Select **Tutorial 1 _sim1.sim** and click **OK**.
3. On the **Simulation Navigator**, expand the Tutorial 1_fem1 node and uncheck the **3D Collectors** node. The mesh is turned OFF.
4. On the ribbon, click **Home > Context > Activate Simulation** .
5. On the ribbon, click **Home > Loads and Constraints > Load Type > Force** .
6. Select the holes, as shown.

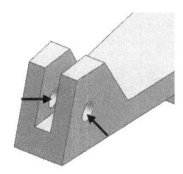

7. Under the **Magnitude** section, type **2000** in the Force box.
8. Under the **Direction** section, click **Specify Vector** and select the Z-axis from the triad.
9. Click the **Reverse Direction** button.
10. Click **OK** to apply the Force load.
11. On the **Simulation Navigator**, expand the **Load Container** node, right click on **Force 1**, and select **Edit Display**.
12. On the **Boundary Condition Display** dialog, drag the **Scale slider** to reduce the size of the load arrows.

13. Click **OK**.
14. On the ribbon, click **Home > Loads and Constraints > Constraint Type > Fixed Constraint** .
15. Select the back face of the geometry and click **OK**.

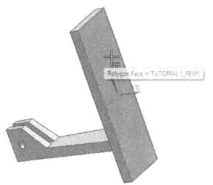

Simulating the Model

Now, you need to check whether the simulation model is setup properly.

1. On the ribbon, click **Home > Checks and Information > More > Model Setup** .
2. Leave all the options checked on the **Model Setup** dialog and click **OK**.

The program checks for any errors during the model setup and displays them in the **Information** window. Also, the Solution-Based Errors Summary displays the following information.

```
Solution-Based Errors Summary
-------------------------------

Iterative Solver Option
More than 80 percent of the elements in
this model are 3D elements.
It is therefore recommended that you
turn ON the Element Iterative Solver in
the "Edit
Solution" dialog.
```

3. Close the **Information** window.
4. On the **Simulation Navigator**, right click on **Solution1** node and select **Edit**.
5. On the **Solution** dialog, check the **Element Iterative Solver** option, and click **OK**.
6. Click **Save** on the Quick Access Toolbar.
7. On the ribbon, click **Home > Solution > Solve**

8. Click **OK** on the **Solve** dialog.
9. Close the **Information** window, **Solution Monitor**, and click **Cancel** on the **Analysis Job Monitor** dialog.
10. On the ribbon, click **Home > Context > Change Display Part > Open Results** .
11. On the **Post Processing** Navigator, go to **Solution 1 > Structural > Stress - Element-Nodal**.
12. Double-click on **Von-Mises**. The result will appear.

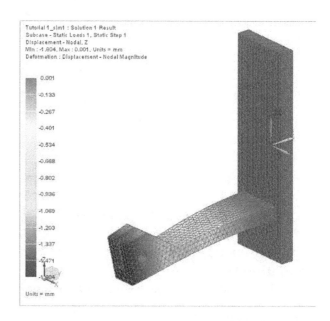

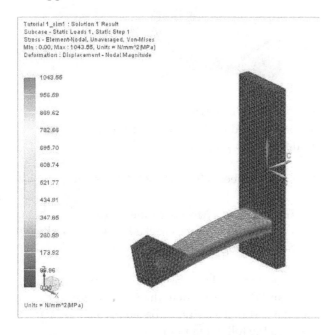

17. On the ribbon, click **Results > Context > Return to Home** .
18. Click **File > Close > All Parts**.
19. Click **Yes Save and Close**.
20. Click **Yes**.

13. On the ribbon, click **Results > Animation > Play** . The model is simulated in the graphics window.
14. Click **Stop** on the **Animation** group.
15. On the **Post Processing** Navigator, expand **Solution 1 > Structural > Displacement – Nodal**.
16. Double-click on **Z**. The result will appear.

Chapter 12: Product and Manufacturing Information

Providing dimensions and annotation in 2D drawings is common and well known method. However, you can provide product and manufacturing information to 3D models as well. The PMI tools help you to add this information to the 3D models based on the universal standards such as ASME Y14.5 – 2009 and ISO 1101: 1983.

In this chapter, you will learn to use **PMI** tools to add GD&T information to parts. The **PMI** tools are available on the **PMI** tab of the ribbon. If the **PMI** tab is not displayed by default, you can add it to the ribbon. Click the **Application** tab on the ribbon, and then click the **PMI** icon on the **Design** group.

TUTORIAL 1
In this tutorial, you will add PMI dimensions and annotations to part model.

1. Download the Tutorial 1 part file from the companion website.
2. Open the Tutorial 1 part file.
3. On the CommandManager, click **PMI > Annotation > Datum Feature Symbol** ⬚.
4. On the **Datum Feature Symbol** dialog, type **A** in the **Letter** box under the **Datum Identifier** section.
5. On the **Datum Feature Symbol** dialog, under the **Origin** section, expand the **Orientation** section.
6. Select **Plane > YC-ZC Plane**.
7. Expand the **Leader** section, and then click **Select Terminating Object**.
8. Select the front face of the geometry.
9. Move the pointer and click to position the datum.

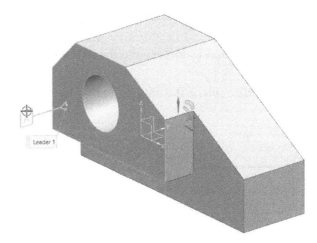

10. Select **Orientation > Plane > XC - ZC Plane**.
11. Click **Leader > Select Terminating Object**.
12. Select the top face of the model.
13. Move the pointer upward, and click to position the **B** datum.

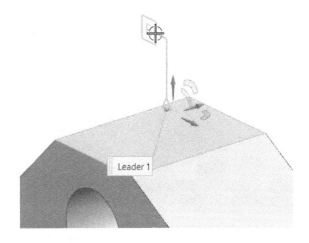

14. Click **Leader > Select Terminating Object**.
15. Select the right face of the model.
16. Move the pointer toward right and click to position the C datum.

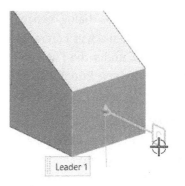

17. Click **Close** on the **Datum Feature Symbol** dialog.
18. In the Part Navigator, expand the PMI node.

Notice the three plane features in the PMI tree.

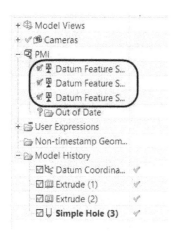

You can click and drag the datum feature sysmbol.

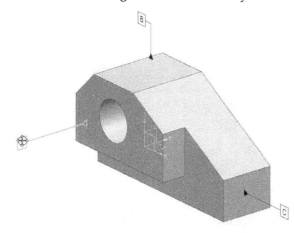

Adding Radial Dimensions

1. On the **File** tab of the ribbon, click **Preferences > Drafting**.
2. On the **Drafting Preferences** dialog, expand the **Dimension** node, and then select **Tolerance**.
3. On the **Tolerance** page, under the **Type and Values** section, select **Type > Equal Bilateral Tolerance** $^\pm X$.
4. Type **2** in the **Decimal Places** box.
5. Type **0.25** and **-0.25** in the **Upper Limit** and **Lower Limit** boxes, respectively.
6. Expand the **Text** node, and then select **Units**.

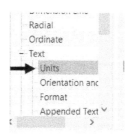

7. On the **Units** page, select **Decimals Delimiter > Period**.
8. Click **OK** on the **Drafting Preferences** dialog.
9. On the ribbon, click **PMI > Dimension > Radial**.
10. Select the hole and place the dimension, as shown.

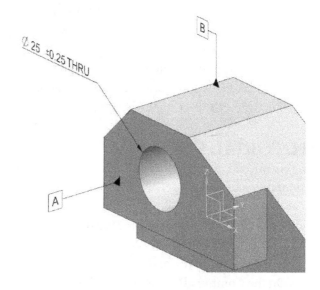

11. Click **Close** on the **Radial Dimension** dialog.

Adding Linear Dimensions

1. On the ribbon, click **PMI > Dimension > Linear**.
2. Select the lower horizontal edge of the geometry.
3. On the **Linear Dimension** dialog, expand the **Orientation** section, and then select **Plane > XC-ZC Plane**.
4. Move the pointer downward and click to position the dimension.
5. Click **Close** on the dialog.

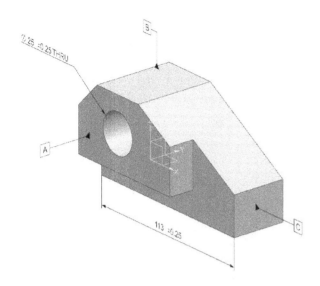

Adding Chamfer Dimensions

1. On the ribbon, click **PMI > Dimension > Chamfer** .
2. Rotate the model and click on the chamfer.
3. Click on the top face of the model to define the reference face of the chamfer (the face from which the chamfer angle is measured).
4. On the Dimension palette, select the **Leader Perpendicular to Chamfer** option from the drop-down available at the bottom.

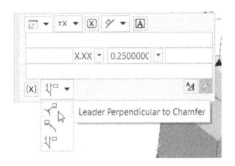

5. Move the pointer and click to position the chamfer dimension.
6. Click **Close** on the **Chamfer Dimension** dialog.

Adding Angular Dimensions

1. On the ribbon, click **PMI > Dimension > Angular** .
2. On the **Angular Dimension** dialog, select **Selection Mode > Objects**.
3. Select the angled and horizontal faces, as shown.
4. Move the pointer toward right, and then click to place the angular dimension.

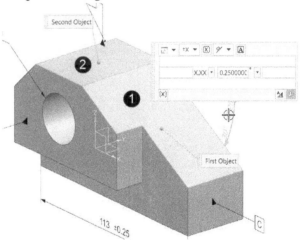

5. Click **Close** on the **Angular Dimensions** dialog.

Using the Rapid command

1. On the ribbon, click **PMI > Dimension > Rapid** .
2. Click the **Reset** icon on the **Rapid Dimension** dialog.
3. Click on the two faces of the model, as shown.

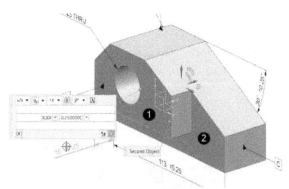

4. On the **Rapid Dimension** dialog, select **Origin >
    Orientation > Plane > YC – ZC Plane**.
5. Move the pointer downward and click to
    position the dimension, as shown.

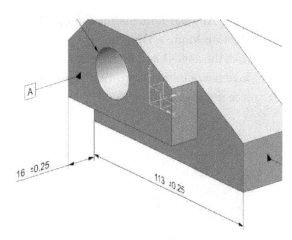

Adding Thickness Dimensions

1. On the ribbon, click **PMI > Dimension>
    Thickness** .
2. Click **OK** on the **Annotation Plane** message box.
3. On the **Thickness Dimension** dialog, click the
    Reset icon.
4. Select the two edges of the model, as shown.

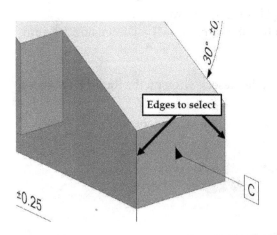

5. On the **Thickness Dimension** dialog, select
    Orientation > Plane > User Defined.
6. Select the face of the model, as shown; the
    dimension is placed on the orientation plane.

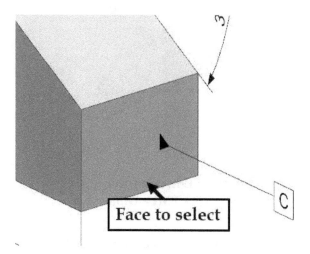

7. Move the pointer downward and click to
    position the dimension.

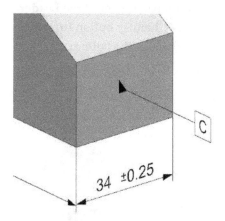

Adding Feature Control Frame

1. On the ribbon, click **PMI > Annotation >
    Feature Control Frame** .
2. On the **Feature Control Frame** dialog, click the
    Reset icon.
3. On the **Feature Control Frame** dialog, under the
    Frame section, select **Characteristic > Position**
    .
4. In the **Tolerance** section, set the values, as
    shown.

5. Under the **Primary Datum Reference** section, select **A** from the drop-down.
6. Expand the **Secondary Datum Reference** section, and then select **B** from the drop-down.
7. Expand the **Tertiary Datum Reference** section, and then select **C** from the drop-down.
8. In the graphics window, place the pointer on the bottom edge of the radial dimension of the hole, as shown; the **Feature Control Frame** snaps to the dimension and a dashed box appears.

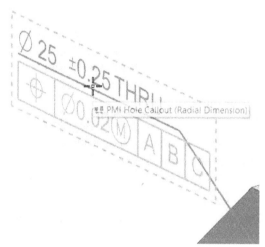

9. Click to place the **Feature Control Frame**.

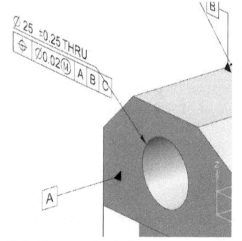

10. Click **Close** on the dialog.
11. Activate the **Feature Control Frame** command.

12. Click the **Reset** icon on the **Feature Control Frame** dialog.
13. On the **Feature Control Frame** dialog, under the **Frame** section, select **Characteristic > Flatness** ⬭.
14. Type 0.02 in the **Tolerance** box.

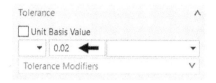

15. Select **Origin > Orientation > Plane > User Defined**.
16. Select the front face of the model.
17. Click **Leader > Select Terminating Object**.
18. On the Top Border Bar, make sure that the **Point on Face** icon is active.
19. Click on the front face of the model.
20. Move the pointer and click to specify the location of the Feature Control Frame.

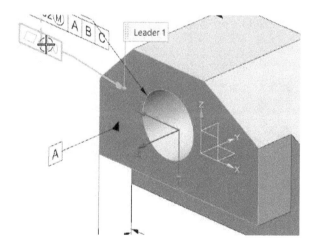

21. Add two feature control frames to the faces, as shown.

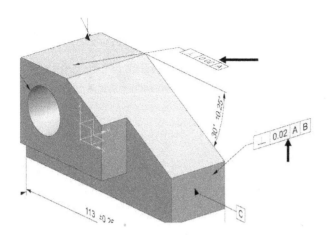

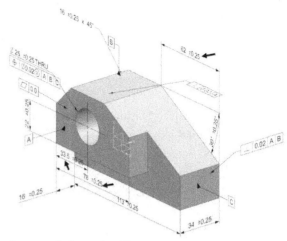

Adding centermarks

1. On the ribbon, click **PMI > Supplemental Geometry > Center Mark** ⬜.
2. Click the **Reset** ⬜ icon on the **Center Mark** dialog.
3. Select the hole from the model.
4. On the **Center Mark** dialog, expand the **Orientation** section, and then select **Plane > User Defined**.
5. Click **Specify CSYS**.
6. Select the front face of the model.
7. Click **OK** on the dialog to create the center mark.

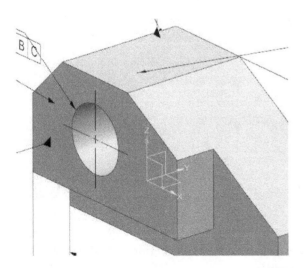

8. Add remaining dimensions to the part.

9. Save and close the file.

TUTORIAL 2 (Using the PMI data in drawings)

1. Open the Tutorial 1 part file.
2. On the ribbon, click **File** tab > **New**.
3. Click the **Drawing** tab on the **New** dialog.
4. Select **A3 - Size** from the **Templates** section, and then click **OK**.
5. Click **Close** on the **Populate Title Block** dialog.
6. On the **View Creation Wizard,** click the **Inherit PMI** option located at the left side.
7. On the **Inherit PMI** page, select **Aligned to Drawing (Entire Part)**.
8. Check the **Inherit PMI onto Drawing** option.
9. Click **Next**.
10. On the **Orientation** page, select **Front** from the **Model Views** list.
11. Click **Next**.
12. On the **Layout** page, select the **Top** and **Right** views.

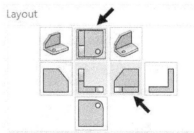

13. Under the **Placement** section, select **Option > Manual**.
14. Move the pointer and click it at the location, as shown.

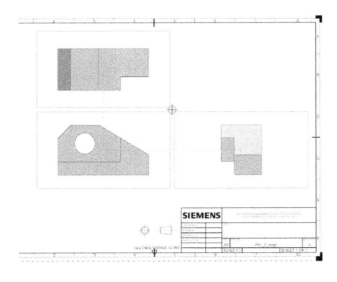

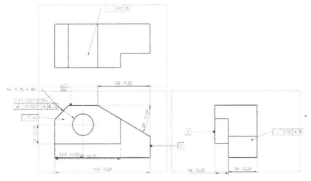

15. Adjust the dimensions by dragging them.

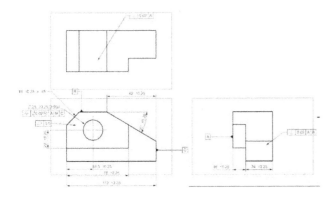

16. Save and close the drawing file.

Chapter 13: Visualization and Rendering

In this chapter, you will:

- Add Materials to the model
- Apply Backgrounds and Scenes to the model
- Render Images
- Add Cameras and Lights
- Add Decals

TUTORIAL 1 (Working in the True Shading environment)

1. Download the Visualization and Rendering part files from the Companion website and open the Tutorial 1 file.

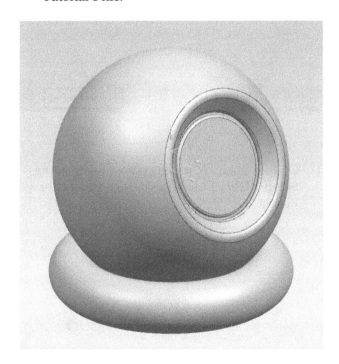

2. On the ribbon, click **View** tab > **True Shading**

You can also right click in the graphics window and select **True Shading**.

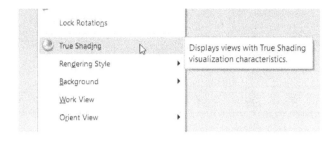

3. On the ribbon, click **View** tab > **True Shading Setup** > **Global Materials** gallery > **Global Material Low Sheen Plastic Wash**.

4. Click on the **Show Shadows** icon to disable it.
5. On the Top Border Bar, select **Type Filter > Face**.

6. Click on the flat face and edge fillet of the model, as shown.

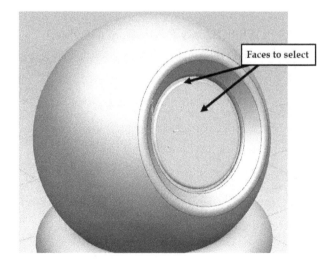

Faces to select

7. On the ribbon, click **View** tab > **True Shading Setup** > **Object Materials** gallery > **White Glossy Plastic**.

8. Click on the edge fillet, as shown.

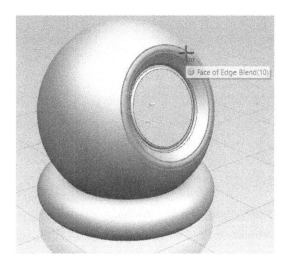

9. Select **White Glossy Plastic** from the **Object Materials** gallery.

10. Select the faces of the model, as shown.

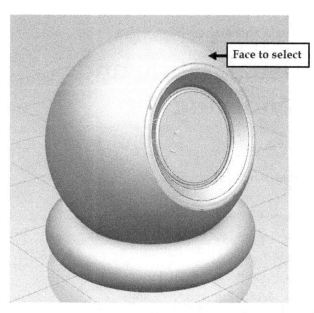

Face to select

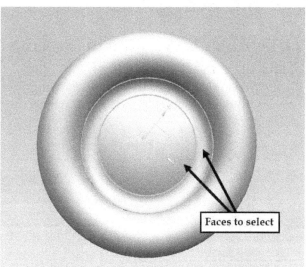

Faces to select

11. On the ribbon, click **View tab > True Shading Setup > Object Materials** gallery **> Red Glossy Plastic**.

12. Likewise, apply **Black Glossy Plastic** material to the remaining faces

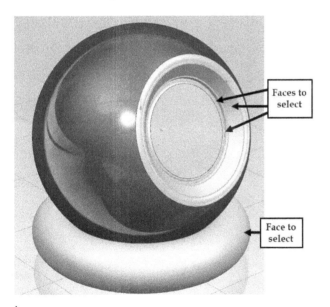

13. On the ribbon, click **View** tab > **True Shading Setup > Background** drop-down > **Inherit Shaded Background**.

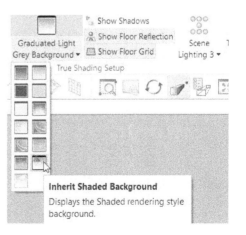

14. On the ribbon, click **View** tab > **True Shading Setup** > **True Shading Editor** 🌑 ; the **True Shading Editor** dialog pops up on the screen.

On this dialog, you set all the options related to the True Shading environment. The **Global Reflections** section has options to display the material with different reflections.

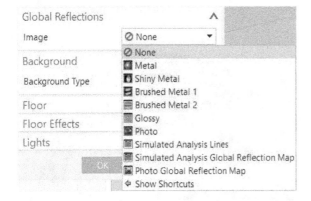

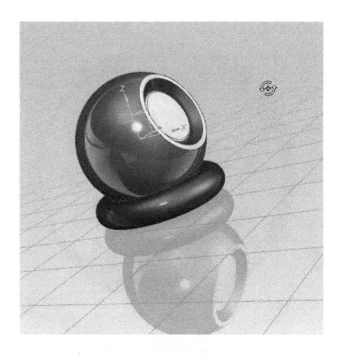

The **Back** ◐ option displays the shadow of the model on a fixed back plane.

The **Floor** section has the options to modify the environment surface. The **Orientation** drop-down has the options to change the orientation of the environment surface to **Bottom**, **Back**, or **Bottom Fixed**.

The **Bottom** 👤 option displays a floating environment surface at the bottom. When you rotate the model, the environment surface changes its orientation along with the model.

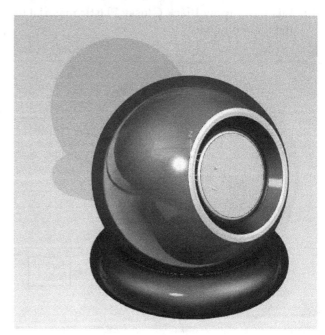

The **Bottom Fixed** 👤 option displays the shadow and reflection of the model on a fixed bottom plane.

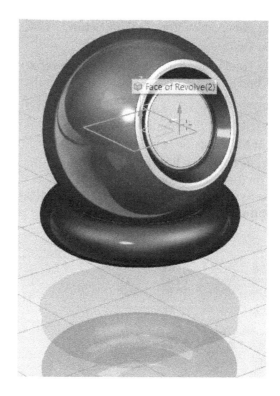

The **Specify Plane** option helps you to change the orientation of the environment surface by creating a new plane.

15. On the **True Shading Editor** dialog, under the **Background** section, select **Background Type > White**.
16. Expand the **Lights** section, and then drag the **Brightness** slider to change the brightness of the lights.

The **Active Scene Lights** drop-down has five different settings.

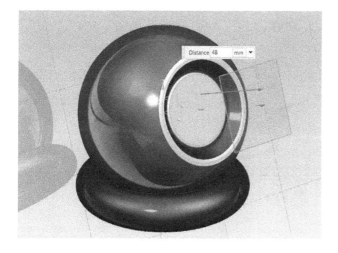

The **Offset** slider helps you to offset the environment surface.

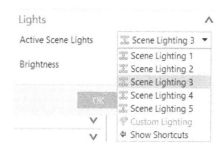

17. Click **OK** on the dialog.
18. Save and close the part file.

Tutorial 2 (Applying Materials in the Studio Task Environment)

1. Download the Visualization and Rendering part files from the Companion website and open the Tutorial 2 file.

2. On the ribbon, click **Render** tab > **Setup** > **Studio Task** .

3. On the Top Border Bar, select the **Type Filter > Solid Body**.

Menu ▾ | Solid Body ▾

4. On the ribbon, click **Home > Studio Setup > System Materials** ; the **System Studio Materials** tab of the **Resource Bar** appears.

5. Click on the model geometry.

6. On the **System Studio Materials** tab, click **Metal > Platinum**.

The selected material is added to the whole part.

Editing the Material

1. Click on the model, and select **Edit Material**.

2. On the **Studio Material Editor** dialog, under the **Structure** section, click **Base Platinum**; the **Properties** section appears on the **Studio Material Editor** dialog.

3. Under the **Properties** section, click **Base type > Silver**.

4. Type **Silver** in the **Name** box under the **Description** section.

5. Click **OK** on the dialog to change the material to silver.

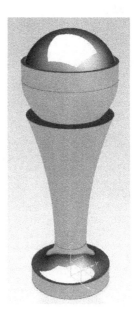

Also, notice the **Silver** material on the **Studio Materials in Part** tab of the **Resource Bar**.

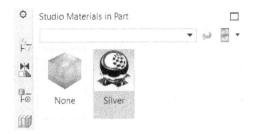

6. On the Top Border Bar, click **Type Filter > Face**.
7. Click on the faces of the model, as shown.

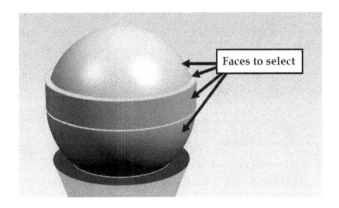

Faces to select

8. On the **System Studio Materials** tab, click **Metal > Brass Cartridge**.

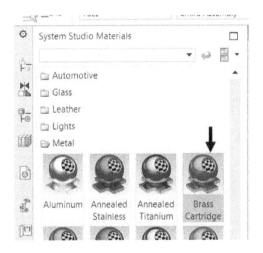

9. Click the **Studio Materials in Part** tab on the Resource Bar.
10. Click the right mouse button on the **Brass Cartridge** material, and then select **Edit**.

11. Type **Gold** in the **Name** box in the **Description** section.
12. Select **Base – MetalBrushed** from the **Structure** section.
13. Under the **Properties** section, select **Base Type > Gold**.
14. Under the **Roughness** section, select **Type > Scalar**.

Place the pointer on the **Type** drop-down and notice the illustration. It shows the level of smoothness from the value that enter in the **Value** box.

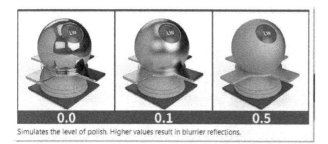

Simulates the level of polish. Higher values result in blurrier reflections.

15. Click **OK**.

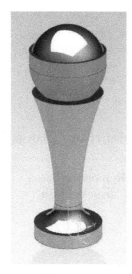

16. Click **Finish Studio** on the ribbon.
17. Save and close the part file.

Tutorial 3 (Scenes)

After adding materials, a Scene is an important part of the rendering process. A material is added to the geometry, whereas a Scene is added to the NX environment. A Scene creates a realistic environment in which the model is placed. For example, you can add a road scene to a car model.

1. Open the Tutorial 3 file.

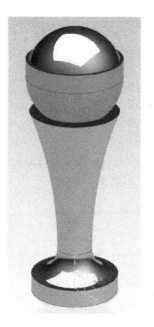

2. On the Top Border Bar, click **Orient View** Drop-down > **Trimetric**.
3. On the Resource Bar, click the **System Scenes** tab.
4. On the **System Scenes** tab, notice the three folders: **Indoor**, **Outdoor**, and **Studio**.
5. Click on the **Indoor** folder and notice the thumbnails displayed in the Resource Bar.

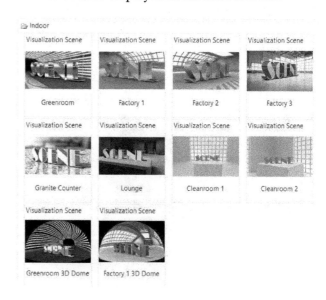

6. Double click on the **Lounge** scene from the Resource Bar.
7. Click the middle mouse button and drag the mouse. Notice that the background remains static.
8. Select the **Factory 1 3D Dome** scene.

9. Rotate the model and notice that the environment rotates along with the model.

10. Likewise, examine the **Outdoor** and **Studio** folders.

11. Close the file without saving it.

Customizing a Scene

1. Open the Tutorial 2 part file.
2. On the ribbon, click **Render > Studio Setup > Scene Editor** .
3. On the **Scene Editor** dialog, click the **Background** tab.
4. Select the **2D Background** from the **Settings** section.
5. Select **Type > Image File**.
6. Click the **Choose from Image Palette** icon.
7. On the **Environment Image Palette** dialog, click the **Indoor** tab.
8. Select the third image from the top.
9. Click **OK**.
10. Click the **Lights** tab and click **Specify Orientation**.
11. On the Top Border Bar, click the **Orient View drop-down > Top**.
12. Click and drag the Angle handle of the Dynamic Coordinate System such that the Light Direction points to the model.

13. Use the **Zoom**, **Pan**, **Fit**, **Orient View**, and **Rotate** tools to fit the model in the background image.

14. Drag the **Intensity** slider to adjust the light intensity.
15. Click **OK**.
16. Click **Finish Studio** on the ribbon.
17. Save and close the part file.

Tutorial 4

1. Open the Tutorial 4 file.

2. On the ribbon, click **Render > Setup > Studio Task**.
3. On the Resource Bar, click the **System Studio Materials** tab.
4. On the **System Studio Materials** tab, click the **Automotive** folder.
5. Select the solid body from the graphics window.
6. Click **Wheel Aluminum** under the **Automotive** folder.

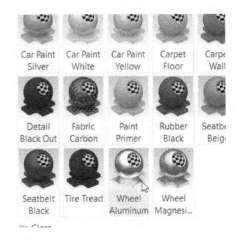

7. On the Resource Bar, click the **System Scenes** tab.
8. Click the **Studio** folder and select the **White Studio**.

Visualization Scene

White Studio

Creating Cameras

1. On the Resource Bar, click the **Part Navigator** tab.
2. Press and hold the Shift key and select all the points and lines available in the Part Navigator.
3. Right click and select **Show**.

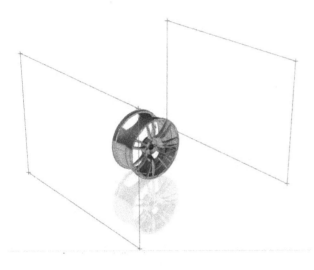

4. On the Top Border Bar, click **Orient View** drop-down > **Trimetric**.
5. In the Part Navigator, click the right mouse button on the **Cameras** node, and then select **Create**.

The **Camera** dialog appears. Also, the **Rendering Style** of the model changes to **Static Wireframe**. The resultant view is displayed in the upper right corner of the graphics window.

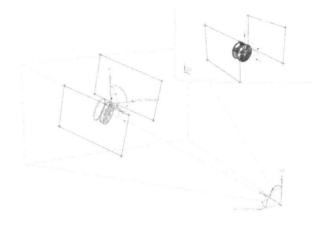

6. Right click in the graphics window, and then select **Orient View > Isometric**.

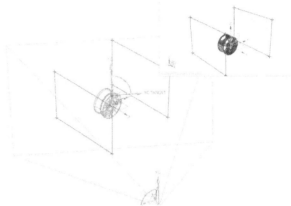

7. On the **Camera** dialog, select **Type > Perspective**.
8. Type **Position1** in the **Camera Name** box under the **Name** section.
9. Click on the origin point of the camera, and then drag. Notice that the target point of the camera is fixed.

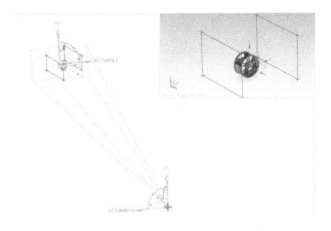

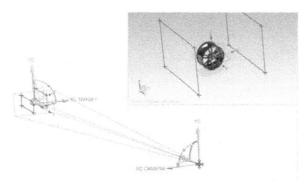

10. Uncheck the **Lock Target Position** option in the **Camera** section.
11. Now, click and drag the camera. Notice that the target point moves along with the camera.

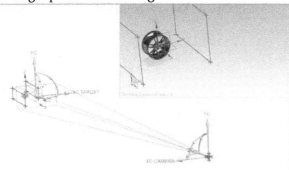

12. Likewise, click and drag the target position of the camera. Notice that the camera position is fixed. You can unlock the camera position by unchecking the **Lock Camera Position** option in the **Target** section.

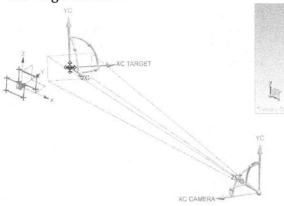

13. Click the **Point Dialog** icon in the **Target** section.
14. On the **Point** dialog, under the **Output Coordinates** section, select **Reference > WCS**.
15. Click the **Reset** icon on the **Point** dialog; the X, Y, Z values of the point are set to 0,0,0.
16. Click **OK** on the **Point** dialog.

17. Check the **Lock Target Position** and **Lock Camera Position** options.
18. Click the **Specify Orientation** option in the **Camera** section.
19. Select the point from the graphics window, as shown.

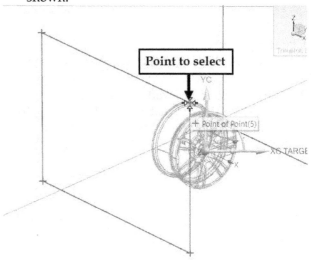

The view is updated in the window displayed at the upper right corner.

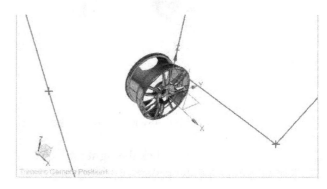

20. Expand the **Custom Zoom** section, and the type 0.14 in the **Magnification** box.
21. Click **OK** on the **Camera** dialog.

Position 3

In the Part Navigator tab, expand the **Cameras** node and notice that the **Position1** camera view is added to list.

22. Likewise, create two more camera views at the positions, as shown.

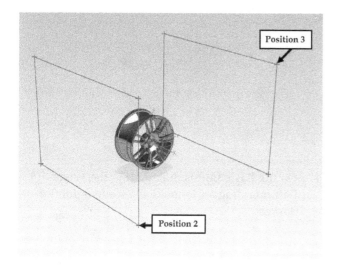

Position 2

Rendering in Ray Traced Studio

1. In the Part Navigator, expand the **Cameras** node.
2. Double-click on the **Position2** camera to activate it.
3. On the ribbon, click **Render >Display > Ray Traced Studio** ; the **Ray Traced Studio** window appears.
4. Click the **Pause** button at the upper left corner to pause the rendering.
5. Click the **Ray Traced Studio Editor** icon.
6. On the **Ray Traced Studio** dialog, under the **Dynamic** section, select **Render > Photoreal**.
7. Set the **File Save Format** to **JPEG**.
8. Set the **Units** to **Pixels**.
9. Set the **Size** to **Use Defined**.
10. Set the **Orientation** to **Landscape**.
11. Select **Resolution > High**. Note that the processing time will increase as you increase the resolution.
12. Leave the other default options and click **OK**.
13. On the **Ray Traced Studio** window, click the **Start/Resume** icon. Notice the preview of the rendering.

14. Click and drag the **Brightness** slider to change the brightness of the rendered image.
15. Zoom in or out using the mouse scroll wheel. Notice that the rendering starts from the beginning.
16. Double-click on the **Positon2** camera in the Part Navigator.
17. On the **Ray Traced Studio** window, click the **Start Static Image** icon. This option renders a high quality image at the current camera position. The rendering will continue even if you change the orientation or camera position in the graphics window.
18. Wait until the image is rendered.
19. Click the **Save Image** icon to save the image.

20. On the **Save Image** dialog, click the **Browse** icon and specify the location of the image file.
21. Click **OK** on the **Save Image** dialog.
22. Close the **Ray Traced Studio** window.

23. Click **Finish Studio** on the ribbon.
24. Save and close the part file.

Tutorial 5

Lights are very important in rendering a photorealistic image. They can enhance the image to look more realistic by highlighting some portions and creating shadows. You can access the lights on the **Basic Lights** and **Advanced Lights** dialogs.

Working with Basic Lights dialog

1. Open the Tutorial 5 file.
2. On the Top Border Bar, click **Menu > View > Visualization > Basic Lights** .
3. On the **Basic Lights** dialog, click the **Reset to Scene Lights** button; all the lights on the **Basic Lights** dialogs are turned on.

4. On the **Basic Lights** dialog, drag the **Scene Dimmer** dragger to make the scene lighter or darker.

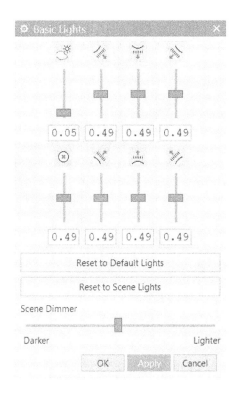

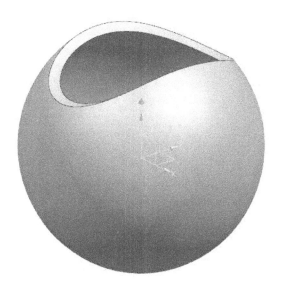

5. On the **Basic Lights** dialog, click the **Reset to Default Lights** button. Notice that only three lights are turn on: **Scene Ambient**, **Scene Left top**, and **Scene Right top**.

6. Drag the sliders of the lights to change their intensity.
7. Click **OK** on the dialog to apply the changes.

Tutorial 6 (Working with directional Lights)

NX adds directional lights to the model, automatically. It is used to highlight a portion of the geometry and display shadows. You can add, delete, move, or adjust a directional light.

1. Open the Tutorial 6 file.

2. On the Top Border Bar, click **Menu > View > Visualization > Advanced Lights** ⮞ .
3. On the **Advanced Lights** dialog, click the **Scene Ambient** ⛭ icon in the **On** section. The ambient light falls on the model from all directions.
4. Click the **Turn Light OFF** ⬇ icon.

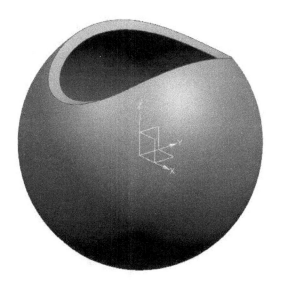

5. Likewise, turn OFF the **Scene Left Top** and **Scene Right Top** lights on the **Advanced Lights** dialog.

Note that the directional lights are also turn ON or OFF based on the type of scene.

6. Click **Cancel** on the **Advanced Lights** dialog.
7. On the ribbon, click **Render > Setup > Studio Task** .
8. On the ribbon, click **Home > Studio Setup > System Scenes** .
9. On the **System Scenes** tab of Resource Bar, select the **Studio** folder, and then select the **White Studio** scene.
10. On the Top Border Bar, click **Menu > Studio > Advanced Lights** .
11. On the **Advanced Lights** dialog, under the **Orient Light** section, select **Show > All Lights**. Notice two directional lights related to the selected scene.

12. Select the **Scene Front** ⊗ light from the **Off** section.

13. Click the **Turn light ON** ⬆ icon.

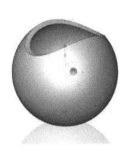

14. Select the **Scene Front** ⊗ light from the **On** section.
15. On the **Advanced Lights** dialog, under the **Basic Settings** section, click the **Color** swatch.
16. On the **Color** dialog, select the red color, and then click **OK**.
17. Drag the **Intensity** slider to 1.0 position.
18. Press and hold the middle mouse button, and drag the mouse; the model view rotates. Notice that the **Scene Front** light also rotates along with the model.

19. Click **OK** on the **Advanced Lights** dialog.
20. On the ribbon, click **Home > Display > Ray Traced Studio**. Notice that the image is rendered without the **Scene Front** light.

21. Close the **Ray Traced Studio** window.
22. On the ribbon, click **Studio Setup > Scene Editor** .
23. On the **Scene Editor** dialog, click the **Lights** tab.
24. On the **Lights** tab, under the **Lights in Scene** section, select the **Scene Front** light.
25. Under the **Light Settings** section, check the **Use with Ray Traced Image-based Lighting** option.
26. Click **OK**.
27. On the ribbon, click **Home > Display > Ray Traced Studio**. Notice that the image is rendered with the **Scene Front** light.

28. Close the **Ray Traced Studio** window.
29. Click **Finish Studio** on the ribbon.
30. Close the part file without saving it.

TUTORIAL 7 (Decal Sticker)

NX allows you to add images such as logos to your model by using the **Decal Sticker** command. Adding decals makes the rendered model look more realistic.

Adding a Decal sticker and embossing it

1. Open the Tutorial 7 part file.
2. Download the decal_sticker.jpg from the companion website.
3. On the ribbon, click **Render > Studio Task**.
4. On the ribbon, click **Home > Studio Setup > Decal Sticker** .
5. On the **Decal Sticker** dialog, under the **Image** section, click the **Choose Image File** icon.
6. Go to the location of the decal_sticker.jpg file and double-click on it.
7. On the **Decal Sticker** dialog, under the Image section, check the **Preview** option.
8. Under the **Object to Decal** section, click **Select Object**, and then click on the face, as shown.

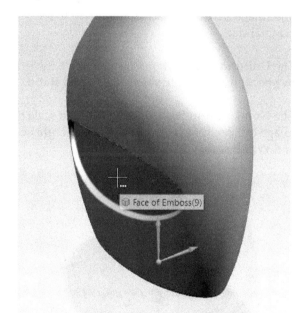

9. Right click in the graphics window and select **Orient View > Front**.
10. On the **Decal Sticker** dialog, under the **Placement** section, select **Anchor Type > Center**.
11. Click the drop-down next to the **Specify Origin** option, and then select **Point on Face**.
12. Click on the approximate center of the emboss face.

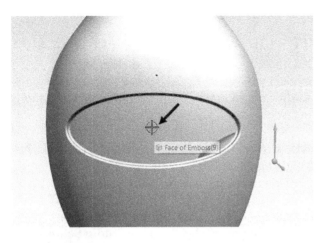

13. Use the **Specify Up Vector**, **Specify Normal Vector** and **Rotation Angle** slider, if required (leave the default options for this example).
14. Under the **Scaling** section, select **Scaling Method > Face Size**.

The remaining options in the **Scaling Method** drop-down are **Image Size**, **Uniform Scale**, and **Non-Uniform Scale**.

The **Image Size** option scales the image to its original size.

The **Uniform Scale** option is used to scale the image by specifying the **Scale** value and **Aspect Ratio**.

The **Non-Uniform Scale** option is used to scale the image by specifying the **Aspect Ratio**, **Height scale**, and **Width scale** values.

15. On the **Decal Sticker** dialog, under the **Transparency** section, set the **RGB Tolerance** value to **50**; the edges of the image are refined.
16. Under the **Reflectivity** section, select **Type > Plastic**.

17. Under the **Displacement** section, check the **Enable Displacement** option.
18. Set the **Amplitude** value to 0.2
19. Set the **Softness** value to 0.2.
20. Click **OK**.

21. Save and close file.

Index